363.19

Changing The Nature of Nature
Genetically Engineered Food

Martin Teitel and Kimberly A. Wilson
Foreword By Ralph Nadar

First published in Great Britain in 2000 by VISION Paperbacks, division of Satin Publications Ltd.

This book is copyright under the Berne convention. All rights reserved. No part of this publication may be reproduced, stored in a retrieval system, or transmitted in any form or by any means, electronic, mechanical, photocopying, recording or otherwise, without prior written permission of the publisher.

Originally published in the United States by Park Street Press.

©2000 Martin Teitel and Kimberly A. Wilson
ISBN: 1-901250-55-5

VISION Paperbacks
a division of Satin Publications Ltd.
20 Queen Anne Street
London W1M 0AY
UK
email: sheenadewan@compuserve.com

Publisher: Sheena Dewan
Cover image: © 2000 Nickolai Globe
Layout: Justine Hounam
Printed and bound in the UK by Biddles Ltd.

CONTENTS

Acknowledgments ix
Foreword by Ralph Nader xi
Introduction: Hijacked Dinner 1

1 – How Genetic Engineering Works 7
2 – What's in Your Shopping Trolley? 21
3 – You Are What You Eat 45
4 – Your Right to Know 61
5 – Food Fights 77
6 – Fields of Green: Farming and Biotech 87
7 – Crossing Swords with an Angel 105
8 – What the Future Holds 113
9 – The Light at the End of the Tunnel – What You Can Do123

Postscript – Britain in GM Revolt 141
Appendix A: Organic Seed Saving 153
Appendix B: Related Web Sites .157
Appendix C: Organisations 163
Suggested Reading 169
Footnotes 181
Index 191

ACKNOWLEDGMENTS

We wish to recognise the many people who helped us in this project. The Council for Responsible Genetics provided special assistance. Two of CRG's summer interns, Terry L. Baynes of Harvard University and Bernadette M. Lake-Willcutt of Mt. Holyoke College, spent many weeks helping us in every phase of writing this book. During the months we worked on this book our friend and colleague, Sophia Kolehmainen, J.D., who directs CRG's Human Genetics Program, cheerfully picked up a thousand and one tasks so we would be freed up for research and writing. This project would have been impossible without Sophie's steadfast support.

The eminent and busy dozen people on the Council for Responsible Genetics's board provided a wealth of information and help; many of them read the manuscript in whole or in part. It is important to note that while these individuals assisted us in many ways, the responsibility for this manuscript, and any errors that might have inadvertently crept in, rests solely with the authors. The CRG board consists of Phillip Bereano, J.D.; Paul Billings, M.D., Ph.D.; Rev. Colin Gracey, D.Min.; Debra Harry, M.A.; Martha Herbert, M.D., Ph.D.; Ruth Hubbard, Ph.D.; Jonathan King, Ph.D.; Sheldon Krimsky, Ph.D.; Claire Nader, Ph.D.; Stuart Newman, Ph.D.; Devon Peña, Ph.D.; and Doreen

Stabinsky, Ph.D.˙ Many people support the Council for Responsible Genetics by volunteering their time and contributing money. We also receive financial support from several foundations, whose assistance is crucial to our work, including the CS Fund, HKH Foundation, Sun Hill Foundation, Solidago Fund, Safety Systems Foundation, the Philanthropic Collaborative, and the Foundation for Deep Ecology.

We also need to mention the great debt we owe to the many colleagues and activists who are working tirelessly on the issue of genetically engineered food. We are proud to be part of a growing social movement that makes our efforts effective and enjoyable.

Our editor at Park Street Press, Rowan Jacobsen, is, as far as we can tell, a perfect editor to work with, excepting only his penchant to favour losing baseball teams.

Finally, each of us wishes to express our deep gratitude to our families and close friends, who cut us all the slack we needed to get the job done. Special thanks go to Marty's wife, the Rev. Mary Harrington, who wrote the grace for us to include in chapter 7.

FOREWORD

Genetic engineering – of food and other products – has far outrun the science that must be its first governing discipline. Therein lies the peril, the risk, and the foolhardiness. Scientists who do not recognise this chasm may be practicing "corporate science" driven by sales, profits, proprietary secrets, and political influence-peddling.

Good science is open, vigorously peer reviewed, and intolerant of commercial repression as it marches toward empirical truths. The rush of genetically engineered foods is leaving behind three areas of science: (1) ecology, often academically defined as the study of the distribution and abundance of organisms; (2) nutrition-disease dynamics; and (3) basic molecular genetics itself. The scientific understanding of the consequences of genetically altering organisms in ways not found in nature remains poor.

Without commensurate advances in these arenas, the wanton release of genetically engineered products is tantamount to flying blind. The infant science of ecology is under equipped to predict the complex interactions between engineered organisms and extant ones. As for any nutritional effects, our knowledge is also deeply inadequate.

Finally, our crude ability to alter the molecular genetics

of organisms far outstrips our capacity to predict the consequences of these alterations, even at the molecular level. Foreign gene insertions may change the expression of other genes in ways that we cannot foresee.

Moreover, as Martin Teitel and Kimberly Wilson point out in this book, the very techniques used to effect the incorporation of foreign genetic material in traditional food plants may make those genes susceptible to further unwanted exchanges with other organisms. Still, the hubris of genetic engineers soars despite an enormously complex set of unknowns.

Corporate promoters, such as the Monsanto corporation, are racing to be first in their markets. Using crudely limited trial-and-error techniques, they are playing a guessing game with the environment of flora and fauna, with immensely intricate genetic organisms, and with, of course, their customers on farms and in supermarkets. This is why these marketeers cannot answer the many central questions raised in this book. They simply do not have the science yet with which to provide even preliminary answers. These companies are so focused on sales that they view with antagonism an independent, open science positing and testing hypotheses with their corporate data.

Selective corporate engineering, unmindful of the need for a parallel development of our knowledge of consequences, can produce disasters.

Costly errors involving past and current technologies – from motor vehicles to atomic power reactors and their waste products to antibiotic-resistant bacteria – should give us pause.

What are the proven benefits of genetically engineered foods that would offset these multifaceted risks? As the authors point out, genetically modified foods "do not taste better, provide more nutrition, cost less, or look nicer." Why, then, would a person run the risk, however large or small it might be, of using them when safe alternatives are

Foreword

available? If the countercheck of science and scientists has been impeded for the time being by the biotechnology industry, what of other precautionary and oversight forces? On this score the record is also dismal. As the engine of massive research and development [subs] and technology transfers to this industry, the federal government has become the prime aider and abettor. In addition, the government has adopted an abdicating non-regulatory policy toward an industry most likely, as matters now stand, to modify the natural world in the twenty-first century. When it comes to biotechnology, the word in Washington is not regulation; rather it is "guidelines," and even then in the most dilatory and incomplete manner. On August 15, 1999, the Washington Post reported that the "FDA is now five years behind in its promises to develop guidelines" for testing the allergy potential of genetically engineered food. The EPA is similarly negligent. To quote the Post article again, "while the agency has promised to spell out in detail what crop developers should do to ensure that their gene-altered plants won't damage the environment it has failed to do so for the past five years." Post reporter Rick Weiss then cited studies showing adverse effects developing that the industry had not predicted.

The US Department of Agriculture has been handing out tax dollars to commercial corporations, including co-funding the notorious terminator-seed project, in order to protect the intellectual property of biotechnology firms from some farmers. You can expect nothing but continuing boosterism from that corner. The creation of pervasive unknowns affecting billions of people and the planet should invite, at least, a greater assumption of the burden of proof by corporate instigators that their products are safe. Not for this industry. It even opposes disclosing its presence to consumers in the nation's food markets and restaurants. Against repeated opinion polls demanding the labelling of genetically engineered foods, these companies have used their political power over the legislative and executive branches of government to block the consumers' right to know and to choose. This issue could soon become the industry's Achilles' heel.

What about universities and their molecular biologists? Can we expect independent assessments from them? Unfortunately, with few exceptions, they have been compromised by consulting complicities, business partnerships, or fear. Although voices within the Academy are beginning to be heard more often, both directly and through such organisations as the Council for Responsible Genetics, the din of the propaganda, campaign money, media intimidation, and marketing machines is still overwhelming. In 1990 Harvard Medical School graduate and author Michael Crichton warned about the commercialisation of molecular biology without federal regulation, without a coherent government policy, and without watchdogs among scientists themselves. He said, "It is remarkable that nearly every scientist in genetics research is also engaged in the commerce of biotechnology. There are no detached observers." There are more such observers now. The situation is changing. One sign is how often Monsanto has to threaten product defamation lawsuits to silence the media and critics, who, although being advised that such suits would almost certainly fail in court, cannot absorb the expense to get them dismissed. As bio-engineered crops cover ever more millions of acres from their start in 1996, the likelihood of side effects and unintended consequences looms larger. Farmers will realise they were not told enough of the truth. And, as more foods containing genetic organisms from other species enter the market, consumers will see there is no escape other than to fight back and demand an open scientific process and response to persistent questions and miscues, with the burden of proof right on the companies. All this and more

Foreword

is why Genetically Engineered Food: Changing the Nature of Nature is so valuable for enlightening what Judge Learned Hand once described as "the public sentiment." For increasing numbers of people who want to eat, to learn, to think, and to act in concert as the sovereign people they aspire to be, the subject of an ever more wide-ranging bio-engineered food supply must be subjected to a rigorous democratic process. As the ancient Roman adage put it: "Whatever touches all must be decided by all." Food – its economic, cultural, environmental, and political contexts – is one of the ultimate commonwealths. The ownership and control of the seeds of life, through exclusive proprietary technology shielded by privileges and immunities, cannot be permitted in any democracy.

Commonwealths can neither be seized by dogmas of intellectual property nor can they abide the domination of narrow commercial imperatives driven by the lucre and myopia of wealthy short-term merchandisers in giant corporate garb.

Ralph Nader

INTRODUCTION
HIJACKED DINNER

Imagine yourself one morning on a modern jetliner, settling into your seat as the plane taxis toward the active runway. To pass the time you unfold your morning newspaper, and just as the plane's rapidly building acceleration begins to lift the wheels from the ground, your eye catches a front page article mentioning that engineers are beginning a series of tests to determine whether or not the new-model airplane that you are in is safe.

That situation would never happen, you say to yourself. People have more foresight than that. Yet something we entrust our lives to far more often than airplanes – our food supply – is being redesigned faster than any of us realise, and scientists have hardly begun to test the long-term safety of these new foods.

The genetic engineering of our food is the most radical transformation in our diet since the invention of agriculture 10,000 years ago. During these thousands of years, people have used the naturally occurring processes of genetics to gradually shape wild plants into tastier, more nutritious, and more attractive food for all of humanity. Until very recently, these evolved food plants were part of the common heritage of humankind. Food plants have been available to all in conveniently small and storable packets – seeds – for distribution, trade, and warehousing. In fact, selective plant breeding has brought food security, greater nutrition, and increased biodiversity, while at the same time protecting food systems against hard times, like natural or economic disasters.

In the new kind of agriculture, a handful of giant corporations have placed patents on food plants, giving them exclusive control over that food. These transnational corporations have altered the minute life-processes of food plants by removing or adding genetic material in ways quite impossible in nature. And like our nightmare vision of the untested airplane, genetically altered food is being quietly slipped into our markets and supermarkets without proper labels, and without having passed adequate safety tests. Furthermore, genetically engineered food confers no advantage to consumers: it doesn't look better, taste better, cost less, or provide better nutrition. To distinguish this different sort of food from the natural food we have eaten all our lives, people give it different names. Here, we call it GM food, and in the US and Canada a new term is being used "genfood".

While we eat this new kind of food and feed it to our children on a daily basis, independent scientists are just beginning to conduct tests to learn about the food's safety. In fact, a person in the United States shopping in a modern supermarket would find out that most food products contain genetically modified ingredients – but the lack of useful labelling of genetically engineered food keeps this information hidden. Meanwhile, economists are determining if our local and national farming will be hurt by this dramatic change in agriculture, and environmentalists are considering the ecological damage that genetically modified plants may cause. Unfortunately these food crops are already growing on millions of acres all around our world: at the end of the twentieth century enough genetically engineered crops are being grown to cover all of Great Britain plus all of Taiwan, with enough left over to

Introduction

carpet Central Park in New York. With this abrupt agricultural transformation, humanity's food supply is being placed in the hands of a few corporations who practice an unpredictable and dangerous science.

As we eat genetically altered food and read about new safety tests, we may start to realise that we are the unwitting and unwilling guinea pigs in the largest experiment in human history, involving our entire planet's ecosystem, food supply, and the health and very genetic makeup of its inhabitants. Worse yet, results coming in from the first objective tests are not encouraging. Scientists issue cautionary statements almost weekly, ranging from problems with monarch butterflies dying from genetically modified corn pollen to the danger of violent allergic reactions to genes introduced into soy products, as well as experiments showing a variety of actual and suspected health problems for cows fed genetically engineered hormones and the humans who drink their milk. And this doesn't even consider slow-acting problems that might not show up for years or decades. Who decided this was an acceptable risk?

On the economic front, trade wars are starting to break out around the world as the countries that produce genetically modified food seek to force other nations to accept it, even when such modified food provides no benefit to recipient nations and raises all the risks mentioned above. Meanwhile, environmental activists warn of "superweeds" and "superbugs" being created by genes that escape from genetically engineered plants. And the file of court cases grows as people questioning this new technology are sued into silence and as activists around the world demonstrate to express their concerns.

Three features distinguish this new kind of food. First and most important, the food is altered at the genetic level in ways that could never occur naturally. As genes from plants, animals, viruses, and bacteria are merged in novel ways, the normal checks and balances that nature provides to keep biology from running amok are nullified. Exactly how genes work is a topic of enormous complexity and some controversy, so it is difficult if not impossible to predict what will happen when individual combinations of genes are created in ways that have never been seen before – and then released into the environment.

The second novel feature of the revolution in our food is that the food is owned. Not individual sacks of wheat or bushels of potatoes, but entire varieties of plants are now corporate products. In some cases, entire species are owned. The term "monopoly" takes on new power when one imagines a company owning major portions of our food supply – the one thing that every single person now and into the future will always need to buy.

Finally, this new technology is "globalised." This means that local agriculture, carefully adapted to local ecology and tastes over hundreds and thousands of years, must yield to a planetary monoculture enforced by intricate trade agreements and laws. According to these trade treaties, local laws that we have come to rely on for protection must take a back seat to decisions made far away by anonymous officials working in secret.

In the forthcoming chapters of this book, we are going to examine the genetic engineering revolution in our food. We're going to have a non-technical look at genetic engineering and how it works. We're going to see who benefits from genetically engineered food and who loses out. We'll take some time to look at risks to health, the environment, and our economy. We'll also consider some of the wider implications of genetically engineered food, including the ethical and spiritual consequences of owning and altering the substance of life. Finally, we'll spend some time looking at the practical steps each of us can take to preserve the independence and integrity of our food supply and to safeguard our ability to make informed choices about what we feed our children and ourselves.

Biotech's commandeering of our food is widespread but hardly inevitable. Tens of thousands of natural seeds still exist to form the basis of a diverse, healthy, and locally controlled food system in our world. With proper attention from ordinary people, our food supply will be put back into the hands of farmers and food suppliers and all the rest of us – for the sake of our health and our environment, and for the future that we leave to our children's children.

1
HOW GENETIC ENGINEERING WORKS

> You can't cross-breed them [a tree and a carrot] because they're sexually incompatible. But with molecular techniques you can take a gene from a tree and put it directly into a carrot – without having to drag all the other tree genes along with it.[1]
>
> Al Adamson, Calgene Inc.

Daily life would be impossible if we paid attention to the countless subtle and minute processes that make it all work. We turn the key in a car ignition without thinking about the combustion of hydrocarbons under pressure, or for that matter about the tiny, carefully planned pattern of hills and valleys in the edge of the key, or the tightly bound molecules that give the shiny key its strength and rigidity. We just start up the car and drive off.

Similarly, we eat our lunch while chatting with a friend, intent on the conversation as we chew up our sandwich and swallow it, upon which it begins to be broken down into components that our body can use to sustain our lives. While there is rarely a need to attend to these hidden details, we sometimes have to focus on the small things in life when our car won't start or we lose our key ring – or when something worries us about the food that we are eating.

With the abrupt and uninvited introduction of genetically engineered food into our supermarkets and restaurants, many of us are looking more carefully into the food we eat, wondering if it poses a threat. Because we cannot see genes any more than we can see hydrocarbons, we are not able to identify what is new or what might be dangerous just by pulling the bread from the top of our sandwich and looking. Genetics and the engineering of genes takes place in a microscopic realm and is a field where most people need to turn to technical experts for explanation.

Because so many experts seem to be tied to personal gain from their involvement in the new technologies, we have to be careful where to look to find objective and fair information. In this chapter we're going to ask some questions about how the new biotechnologies work, and how agriculture – the process that produces most of our food – is being reshaped by this new set of techniques.

GENETIC ENGINEERING

In genetic engineering of food crops a gene, or piece of DNA from one source (a fish for example), is isolated, removed, and then "pasted" into the DNA of another source, a tomato for example. The DNA that might interest scientists is removed from one living organism by enzymes – proteins that affect the chemical processes inside a plant – and then moved to another living thing, to be rejoined with its host's DNA in new combinations. DNA is in a shape commonly called a "double helix," which if we could see it would look something like a twisted ladder.

Every gene on the twisted strand of DNA carries instructions for the production of a protein. Scientists say that a gene, one piece of information in the DNA, "codes" for a protein. Genes, whether in humans or carrots, do nothing more than produce proteins that combine with other proteins that were coded for by other genes. The interaction of huge numbers of proteins is almost unimaginably complex; the exact mechanisms by which the protein produced by any one gene contributes to a specific change or characteristic is only dimly understood. Because each gene codes for only one protein, it takes a great many genes to produce even a simple living organism. Corn, for example, has about 250,000 different genes, most with distinct functions. Genes are arranged linearly along the DNA molecule, which is packaged into structures called chromosomes. Every cell in a plant carries copies of all the chromosomes of the plant. Natural species barriers make crosses between unlike living things impossible, so genetic engineers have to find ways to smash through these barriers – even if the results are not viable or are dangerous.

In genetic engineering, "foreign" DNA – the DNA that is brought in from another living thing – is carried to the target plant by vectors. Vectors function like the plastic tubes used by drive-up banks to send slips of paper and money back and forth between the bank and the automobile: they are just carriers. In genetic engineering, viruses are commonly used as vectors, because viruses typically attack the host's cells and slip right into the cell's DNA.

Genetic engineers attach a piece of DNA to a viral vector and then insert the vector into the new organism so it can infect the cells of the target organism, thus delivering the new DNA fragment into the DNA of the target. Because all this cutting and pasting at a submicroscopic level is very difficult to keep track of, scientists often mark the vectors with antibiotic-resistant genes so that normal cells can be distinguished from genetically engineered cells. The cells are doused with antibiotics, and those cells that have incorporated the foreign DNA and the resistance genes from the vector grow, while those that haven't been modified die.

In trying to understand how genetics works, it is

important that we not get carried away with the colourful imagery of science writers; genes are not really little switches that can be turned off and on like a stereo or a lamp. While a small number of genes seem to have a direct one-on-one relationship with a specific trait or characteristic, most genes operate in combination with other genes. Moving a gene around may or may not produce the same result each time, because the gene and protein environment can be fantastically complex and infinitely variable. Most genetic engineering is a highly imprecise practice.

Genetic engineering is often called "genetic recombination" because it literally recombines DNA inside a plant. Several different terms are used to refer to the results of genetic engineering, such as recombined, engineered, modified, or manipulated. A crop that has been genetically engineered is often called a "transgenic" crop, meaning it contains genes from different sources.

THE UNCERTAINTIES OF GENETIC ENGINEERING
Genetic engineering has allowed scientists to splice fish genes into tomatoes, to put virus genes in squash, bacterium genes in corn, and human genes in tobacco (to "grow" pharmaceuticals). Plants with these mixtures of genes are already in development and some of them may be on your supermarket shelves. Normally, the boundaries between species are set by nature. Until recently, those biological barriers had never been crossed. Genetic engineering allows these limits to be exceeded – with results that no one can predict. Some people, such as author Michael Pollan, feel that the boundaries of nature are in place for a reason. He writes, "The introduction into a plant of genes transported not only across species but whole phyla means that the wall of that plant's essential identity – its irreducible wildness, you might say – has been breached."[2]

How Genetic Engineering Works

It may seem bizarre – or even offensive, if you are a vegetarian – to think that the tomatoes you select in the supermarket could have fish genes in them. Worse still, some of these combinations might be not only peculiar but downright dangerous to our environment or even our own health. As molecular geneticist Michael Antoniou puts it:

> The artificial nature of GM [genetic modification] does not make it dangerous. It is the imprecise way in which genes are combined and the unpredictability in how the foreign gene will behave in its new host that results in uncertainty. From a basic genetics perspective, GM possesses an unpredictable component that is far greater than the intended change.[3]

It took geneticists over 270 tries to clone the sheep "Dolly." The 270 Dollys that didn't make it? Well, many of them were deformed and disfigured, stillborn, or unable to mature. Genetic engineering techniques also create many abnormal plants in the process of obtaining a few that live and function somewhat as intended. From the plants that survive DNA recombination, genetic engineers develop commercial-grade seeds. Michael Pollan visited a Monsanto laboratory. He reports:

> The whole operation is performed thousands of times, largely because there is so much uncertainty about the outcome. There's no way of telling where in the genome the new DNA will land, and if it winds up in the wrong place, the new gene won't be expressed, or it will be poorly expressed, or the plant might be a freak. I was struck by how the technology could be astoundingly sophisticated while also being a shot in the genetic dark.[4]

Many geneticists do take a "shot" in the genetic dark, and they do it with what some people call the "gene gun." The gene gun literally blasts microscopic gold bullets coated with foreign DNA at plant cells. Sometimes the bullets pass through the cell walls and the DNA becomes part of the plant's genetic structure. Other times the introduced DNA will miss the cells altogether, or kill them, or break through the cell wall and not become part of the cell's genetic structure. Roberto Verloza, an engineer, writes, "Blasting plasmid-coated microparticles into cells to improve plant characteristics is like putting several screws, nuts, bolts, springs, etc. into a shotgun cartridge, then blasting them into a running engine to stop an occasional vibration in the engine. Perhaps, after several million attempts, the vibration actually stops and they'd be proud of their work. As a real engineer, I refuse to call that engineering. That plant genomes can take such treatment and survive is a testament not to the precision of the genetic engineer, but to the resilience of living organisms to abuse."[5] As we will see in chapter 2, many of the plants that survived genetic engineering still did not produce foods which were up to standard; in fact, the industry has had more than its share of commercial failures thus far.

Because genetic recombination is so imprecise, it can sometimes have unintended side effects. For example, "Genes for the colour red placed into petunia flowers not only changed the colour of the petals but also decreased fertility and altered the growth of the roots and leaves. Salmon genetically engineered with a growth hormone gene not only grew too big too fast but also turned green."[6]

What could be the risks of eating GM food? Studies have already found at least one genetically engineered food, a common soybean, that has less nutritional value than its natural counterparts. We are sure to discover other important and potentially dangerous differences as genes are added, deleted, and shuffled inside our food. How else is our food altered, and how do these alterations affect human health?

TEST TUBE FOODS
The genetic engineering of food crops is a new technique that is still in its infancy. Traditional plant breeding and

agricultural practices have a history that stretches back 10,000 years, and the biotechnology industry tries to tell the public that genetically engineered foods are just a natural extension of this process – nothing to be alarmed about.

But genetic engineering is a radical departure from traditional plant breeding. Plant genetic engineering is a hit-or-miss process. Farmers and plant breeders know that genes are mutable, and many factors, including the environment, play a major role in the expression or adaptation of a gene or genetic trait.

Unlike the hype that surrounds new genetically engineered crops, classical plant breeding methods have steadily increased crop yields every year for decades.[7] These techniques include breeding domestic crops with their wild relatives because the wild relatives provide strong survival traits that can improve domesticated crops. By relying on the birds and bees rather than the microscope and petri dish to determine the success or failure of their new crops, classical plant breeders take advantage of nature's vast storehouse of information, accumulated over millions of years of experimentation, as to what works and what doesn't. It may be slow, but it ensures that no catastrophic mistakes are made.

The idea of using biotechnology to synthesise genetic blueprints for food is an exciting one if we believe predictions of drastically improved crop quality and quantity. Proponents of noble scientific advancement couple progress and efficiency with idealistic notions of improving public health, feeding the poor, and saving the environment.

However, such quick-fix thinking opposes all the foundations of sustainable agriculture, such as natural diversity, balance, and ecology. Let's look at some of those issues.

Natural Diversity

The survival of a species in a continually changing natural environment relies fundamentally upon diversity – diversity in circumstances of growth and reproduction, diversity in environmental adaptations, and diversity in genetic material. Although such diversity may require that some individuals in a species be at a disadvantage at some point in time, it prevents complete eradication of the species when there is a flux in environmental conditions. Natural, random processes such as insect or wind pollination enhance diversity, resulting in offspring with new combinations of genetic material. Although some of these new combinations could be failures in a particular climate that is cool and moist, for example, they would be the potential saviours of the species when conditions change to hot and arid. Thus, biodiversity is essential to natural selection and evolution.

Appreciation of the importance of biodiversity dates back a hundred centuries to the beginning of the agricultural process. By observing nature, humans have successfully selected the best types of crops to grow. Farmers remained powerless, however, when it came to the interaction between crops and their environments. No one could predict whether a season would be wet or dry. Consequently, farmers quickly learned the importance of diversity: maintenance of various crops that thrived under a variety of conditions to avoid entire crop failures and starvation. As we will soon see, the diversity we need for survival is threatened by the corporate-oriented monoculture of GM food.

Selective Breeding

Our economy is centred around control and efficiency. Because modern agriculture is a profit-based venture, the existence of failures, inherent to biodiversity, runs counter to our sense of efficient, flawless mass production. With continuing corporate consolidation and rising industrialisation, the evolution of agriculture has been characterised by increasing manipulation and control over crop gene pools. The earliest farmers selected the types of plants to grow. After a while, people discovered vegetative propagation – a process in which, for example, a piece of a bulb can be broken off and replanted to produce a duplicate to the parent plant. Using this technique, people could control the genetic composition of the next generation. Later, during the scientific and industrial revolutions, farmers used their practical knowledge of genetics to breed plants with preferred characteristics by selective saving of seeds. Until recently, selective seed saving was the basis for all of our food production.

Hybridisation
A monumental event in agricultural development and an important step on the path toward genetic engineering was the discovery of "hybrid vigour." Farmers found that some hybrid organisms stemming from the interbreeding of two different purebred strains were particularly robust. The hybrids displayed enhanced yield, vitality, disease resistance, and life span.

Thus, plant breeders extended human or "artificial" selection, crossing distinct but not entirely unrelated strains of organisms of the same species. Initially, hybridisation was performed on a small-scale, trial-and-error basis by individual farmers in an attempt to improve the quality of their crops. Their goal was a high-quality, more reliable food supply based on plants better equipped to deal with unpredictable environmental challenges. They collected the seeds of each successive crop generation, experimented with new crosses, and traded seeds with their neighbours. Farmers did not "see themselves as breeders of specialised seeds for a global market. They wanted good reliable seed for their own food and that of their neighbours and village.

Consequently, seed was traded around and experimented with freely."[8] The nature of hybridisation changed in the mid-1930s as the technology spread from corn to sorghum, wheat, and sunflowers, sparking the rise of a consolidated "seed industry." Powerful agricultural companies including DeKalb and Pioneer Hi-Bred initiated large-scale research programs devoted to hybrid seed development. Modern industrial hybridisation gave rise to economic dependency when farmers, under economic pressures, began to purchase these genetically uniform hybrids each year, abandoning their traditional practices of seed collecting and hybrid experimentation. Along with farmer dependency and centralisation of the global food supply, industrial hybridisation undermined diversity and set a dangerous legal precedent for the corporate patenting of life.

Genetic Engineering:
A Continuation of the Agricultural Tradition?
Proponents of biotechnology have argued that genetic engineering is the natural next step in the process of agricultural evolution that has been developing for thousands of years. Farmers have always altered the natural order of crop growth, first selecting what crops to grow, then breeding for desired traits, and finally crossing distinct strains to create stronger hybrid progeny. Although genetic engineering may seem like the next logical step in crop control – both in breeding high-quality crops and ensuring survival in a mercurial environment – it does not necessarily represent progress.

Like industrial hybridisation, the genetic engineering of food encourages monoculture: the growth of acres of genetically uniform crops. This dismissal of the importance of biodiversity has had catastrophic effects in the past. "The history of modern agriculture is a history full of

How Genetic Engineering Works

disease – and insect – caused crop disasters, many brought on by the use of a few popular crop varieties."[9] For example, the great Irish potato famine that hit Ireland in 1845 was the result of uniform, widespread growth of a particular potato variety that lacked resistance to a potato blight. Furthermore, in 1972, the National Academy of Sciences reported that modern wheat rust epidemics "are clearly genetically based, in that as resistant varieties become available, fungus mutates to a form that attacks the new variety, and an epidemic ensues."[10]

The loss of valuable genetic traits, called genetic erosion, has been a continuing part of the domestication of food plants. There is no question that modern crop varieties would fail miserably in the wild, because these varieties have been bred for agricultural performance: growth rate, size, taste, and colour. Humans have shaped the genetic makeup of these crops through selective breeding methods, causing the plants to lose certain genetic defence mechanisms such as disease resistance.

Plant species in the wild are less vulnerable to insects and diseases because they have a broader genetic base to work from. While such genetic erosion is irreversible, we must guard against a further loss of biodiversity – the diversity genetic engineering threatens most. When insects and microbes ravage monocultures of GM food, genetic engineers become "treadmill plant breeders," continually forced to develop new pest-resistant varieties that survive for a while – until the pests adapt to that variety.

There are fundamental differences between genetic engineering and traditional agriculture. Prior to the advent of genetic engineering, all selection was based on the crossing of similar plants of one variety.

Acting as outside observers, humans noted what traits were most desirable and they shaped natural breeding processes to select for those traits. Farmers, familiar with breeding, cultivation, and seed collection, understood these genetic processes and so carried out their selective breeding to produce plants that fit local conditions and community standards.

With genetic engineering, the farmer is no longer needed to observe and react as a "selector." Just as hybridisation brought about enforced dependency of the farmer, along with corporate consolidation of the food supply, so does genetic engineering separate the farmer even further from the agricultural process.

Genetic engineering also alters the fundamental composition of crops. The dangerous power of such technology is immense. While natural crossbreeding restricts the types of hybrids that can be produced, genetic engineering allows for gene transfers between vastly unrelated species, such as fish and tomatoes. As we shall see later, aggressive gene-transfer mechanisms and the large-scale effects of such genetic manipulation are imprecise and dangerously unpredictable.

When we start to alter the genetic composition of organisms, we take into our own hands the instructions for life – instructions that have been slowly and carefully evolving since the first appearance of life on this planet, instructions that support the delicate balance of our

---

TIMELINE

1953 Watson and Crick describe the DNA double helix
1983 Creation of first genetically engineered plant
1987 First genetically engineered plants grown in field under USDA supervision
1990 First genetically engineered plants grown for commercialisation under USDA supervision
1996 6 million acres of genetically engineered plants grown in United States
1997 25 million acres grown
1998 58 million acres in United States,[11] approximately 70 million acres worldwide[12]

ecosystem. In assuming the immense responsibility to change those basic instructions, we must honestly and thoroughly analyse every possible motivation and ramification of this novel technology – not only environmental, but social, political, ethical, and economic as well.

2
WHAT'S IN YOUR SHOPPING TROLLEY?

> Monsanto should not have to vouchsafe the safety of biotech food. Our interest is in selling as much of it as possible. Assuring its safety is the FDA's job.[1]
> Phil Angell, Director of Corporate Communications, Monsanto

BIG AGRIBUSINESS

As you wander along the aisles of the supermarket, you select a variety of fresh produce to put in your shopping trolley. While you may not know it, choosing any number of vegetables, including eggplant, squash, tomatoes, corn, cabbage, lettuce, and potatoes, puts you at risk for eating genetically engineered foods. Because there is no requirement in the United States to label genetically modified food, there is no way you can tell just by looking.

Worse yet, many of the growers and brand names you have grown to recognise and trust for fresh produce have been bought out by chemical giants such as Monsanto, Dow, and DuPont. These multinational companies have spent billions of dollars acquiring seed companies to grow and sell their genetically engineered seed. However, the smaller seed companies don't change their names or labels; only the food changes. This industry, often referred to as the "agribusiness" industry, includes seed sellers, chemical manufacturers, and the makers of genetically engineered products. This chapter will give you an overview of the types of genetically engineered foods that exist and, later, the inherent risks involved in growing and eating these foods.

Genetically modified food has quite a checkered history. Proponents of genetic engineering see this history of mistakes and failures as mere "glitches" in a sophisticated plan, which they believe will ultimately succeed. To many others, these "glitches" are ominous signs that genetic engineering is neither a safe nor reliable method to increase the quality or quantity of our food supply. As we saw in chapter 1, the technique of genetic engineering is anything but predictable, and there are many lessons to be learned from past mistakes.

FLAVR SAVR AND OTHER FLUNKIES

The first genetically engineered food product to reach the market was Calgene's Flavr Savr tomato. Approved for sale in May 1994 by the Food and Drug Administration (FDA) of the US government, Calgene's genetically engineered tomato ripened for a longer period of time on the vine, supposedly improving its flavour – hence the name Flavr Savr. The tomato was also supposed to maintain firmness during shipping, reducing losses. Calgene's new tomato was created by isolating a gene that codes for an enzyme in the ripening process. By reversing the gene's activity, Calgene blocked the expression of the ripening enzyme, thereby extending the time it took the picked tomatoes to become soft and thus allowing them to spend more time ripening on the vine.

Calgene's tomato had great commercial promise, because extended shelf life allowed for tomato production in far away places where agricultural standards are less stringent than in the United States. In fact, early in the development of the Flavr Savr, Calgene contracted with Mexican growers for land to produce the Flavr Savr tomatoes for US markets using Mexican labour, since the tomatoes were too soft to be harvested mechanically.[2] At the same time that Calgene outlined cheap production methods, it planned to market the tomato in the United States as high-priced gourmet produce.

The commercial failure of the Flavr Savr began almost as soon as it hit the market. Calgene decided to put labels on their tomatoes, informing consumers that the food was genetically engineered. Wary consumers and negative press attention triggered the company's removal of such labels, even though consumers' demand to know exactly what was in their food remained high.[3] Calgene even tried marketing the tomatoes as a gourmet product under the friendly sounding "MacGregor's" brand name, but despite their presence in thousands of US supermarkets, consumers did not want to pay more for genetically engineered tomatoes.[4] Scepticism from the public also sent Calgene's commercial partners scrambling. Campbell's Soup Company, originally a commissioner of Calgene's work, went on record with the statement, "We have no current...plans to market any bio-engineered products."[5] Calgene also ran into production problems. The Flavr Savr grew well in the laboratory but encountered serious obstacles in the field. Calgene had to abandon genetic engineering to solve these practical problems.

Although the company spent about $25 million to create and market the Flavr Savr, it was never able to grow the tomatoes successfully on a large scale. Several critics pointed out that the same delayed ripening effect could have been produced by traditional breeding methods, and the *Wall Street Journal* reported that, "Calgene's peer-reviewed scientific articles on the altered tomato gene don't provide a ringing affirmation of its Flavr Savr technology."[6] Calgene's "miracle" Flavr Savr gene turned out to be just one of many genes related to the ripening process in tomatoes. In addition, although Calgene's tomato was supposed to be easily shipped, the company experienced problems when they tried to harvest and transport their soft, ripe tomatoes in the same rough fashion used for most tomatoes, which are picked and shipped green.[7] Questions about the nutritional value of delay-ripening produce were also raised, because foods generally lose nutritional value as they age.

The Flavr Savr tomato may have looked and tasted like the real thing, but what nutritional value did it have? Even more serious questions were raised about the presence of antibiotic-resistant genes in the Flavr Savr after it was reported that the tomato contained genes conferring resistance for the antibiotics kanamycin and neomycin. Flavr Savr was finally pulled from the market in 1996, the same year that Calgene was partially purchased by Monsanto. Calgene, fully acquired by Monsanto in January 1997, still works to genetically engineer tomatoes and strawberries.[8]

BLIND SCIENCE
Genetic engineers had watched the Flavr Savr failure closely. Although they believed they could overcome these early failures, a whole host of other engineered food crops suffered a similar fate. The primary reason behind this persistence in the face of repeated failure is profit. The fresh produce market in North America has been estimated at $80 billion dollars, so potential payoff on a successful product is huge.[9]

Before any genetically engineered food crop can be grown in the United States, it must pass a field trial. Field trials, which sometimes last as little as ten weeks, are the crucial step between the laboratory and the commercial market. The safety of a crop is not tested in a field trial, only its ability to survive and develop normally under natural conditions. When corporations conduct field trials of genetically engineered plants, they are growing those plants outside the rigorously controlled conditions of the laboratory. Plants that do well in the lab might develop differently, or poorly, in the real world. Unfortunately, experimental plants in the outdoors can have their genetically altered pollen carried away by insects and wind. If scientists decide that the plants are not useful – or discover that they are hazardous – it is too late to recall the genetic material that has dispersed and possibly combined with neighbouring plants. Several early foods didn't make it through field trials, including DNA Plant Technology's Endless Summer tomato, which was designed to have longer shelf life. DNA Plant Technology is now owned by Empresas La Moderna, a Mexican seed company that controls approximately 40 percent of US produce.[10]

The Ecologist magazine reported a number of other biotech failures, including "the transgenic 'Innovator' herbicide-tolerant canola, which failed to perform consistently in Canada. Due to the appearance of an 'unexpected' gene, Monsanto was forced to withdraw two genetically engineered varieties of canola seed from the Canadian market. That is, after already selling 60,000 seed bags across Canada."[11] Why didn't scientists notice the "unexpected" gene in the laboratory or in field trials? Selling first and asking questions later, in the case of "Innovator" canola, seems to be business as usual. If not the scientists, who can predict the outcome of genetically engineered seeds? Monsanto's Innovator canola provides further evidence of the extremely unpredictable nature of transgenic crops. "A number of different viral-resistant transgenic plants engineered with a viral gene actually showed increased propensity to generate new, often super-infectious viruses by recombination."[12]

Using similar genetic engineering methods, genes from an arctic fish that code for an antifreeze protein have been integrated into tomatoes in an attempt to confer resistance to frost.[13] Presumably, frost-resistant tomatoes could retain their texture after being frozen. Another attempt to avert frost damage was the ice-minus bacterium. Developed by Stephen Lindow at the University of California, Berkeley, it was one of the first genetically modified organisms to be released into the agricultural ecosystem. Ice requires a regularly shaped surface on which to crystallize. Using the tools of genetic engineering, Lindow spliced out the gene responsible for giving the bacterium *Pseudomonas syringae* a regular surface. This "ice-minus" bacterium was sprayed on strawberries, where it outcompeted the natural bacteria, covered the plants, and prevented ice from forming on them, thereby avoiding frost damage.[14]

While these transgenic, frost-resistant organisms continue to be produced and used, serious concerns remain regarding the ecological consequences of dispersing transgenes into the environment. Will these genetically modified micro-organisms persist, upsetting the ecosystem? Could genetically engineered organisms such as the ice-minus bacterium disrupt ice formation – a crucial ecological process? Imagine a world in which a lab – created bacterium prevents ice formation from occurring.

BT TOXIN GENES AND PEST RESISTANCE

Without much success in engineering crops to resist common viruses or to have commercially desirable traits such as increased shelf life, fresh appearance, or better texture, many scientists have tried to genetically engineer food crops with pesticide genes. In particular, genetic engineers have been working frantically with genes from a naturally occurring soil bacterium, *Bacillus thuringiensis* (Bt). Here they have had more success.

Pests such as insects, worms, and nematodes can destroy portions of many food crops, and persistent pests can leave a farmer with serious crop damage and loss of income. Although much of our food is produced with the use of chemical pesticides, there are many other farming techniques that control or deter pests. One of the most successful substances used to get rid of crop pests comes not from a chemical factory but from Bt, a bacterium whose proteins are an insecticide. The primary targets of Bt include caterpillars and some beetle and fly larvae, which stop eating, shrivel up, die, and decompose upon consumption of the Bt insecticide.

Organic and environmentally friendly farms have found that Bt can be sprayed as an effective pesticide that is non-toxic to humans and most other non-pest species.[15] Although Bt is "friendly" in that it doesn't pollute groundwater the way other pesticides do, it has proven deadly for ladybirds, lacewings, and monarch butterflies when engineered into plants.[16]

In 1970 commercial geneticists became interested in Bt's safety and efficacy, so they quietly began collecting samples of Bt in order to "map" its genetic structure. When they found that Bt's genetic structure was relatively simple, they began experimenting with ways to genetically engineer crops to produce the Bt insecticide inside each plant. Earlier attempts to engineer bacteria with genes from Bt did not go smoothly. Sheldon Krimsky, an expert in the field of agricultural biotechnology and professor of Urban and Environmental Policy at Tufts University, writes, "Tests had shown that [genetically engineered Bt] bacteria isolated from infected insects were capable of infecting other insects. This could be viewed as a desirable characteristic in a pest control agent since the infection could spread through the pest population. However, it raises the possibility that the recombinant bacteria would colonise wild plant species and kill non-target insects, leading to unintended, undesirable, and perhaps unpredictable and uncontrollable ecological effects."[17] The bacteria Professor Krimsky is referring to were developed by the Monsanto corporation; due to the great risks involved, field tests were never conducted and the research was abandoned.[18] Plants genetically engineered with Bt were thought to be more profitable than bacteria engineered with Bt. Field testing of these genetically engineered crops began in the late 1980s and the crops were first approved by the US Environmental Protection Agency (EPA) in 1995.

The EPA does not usually regulate food products, since that is the job of the FDA. However, in the case of Monsanto's genetically engineered New Leaf potato, the potatoes have high enough levels of pesticide in them that they are regulated by the EPA. In fact, the EPA regulates all genetically modified pesticides (bacteria) and genetically engineered plants containing pesticide genes because of their high levels of toxins, and it has emphasised that Bt falls under regulation like chemical pesticides.[19] By capturing the Bt market, a strong niche in the $250 million bio-pesticide market,[20] companies like DuPont and Monsanto stand to make small fortunes. However, the use of Bt crops will be short-lived.

Genetically engineering a food crop with Bt is very different from spraying traditional crops with Bt. Normal crops are only sprayed with Bt sporadically, as needed, but crops genetically engineered with Bt produce insecticide during the life cycle of the plant. This constant exposure to Bt can and almost certainly will lead insects to develop resistance to it, threatening its effectiveness for both the genetically engineered crops and organic farmers, as well as other growers.

Margaret Mellon of the Union of Concerned Scientists writes that "widespread use of Bt crops could lead to the loss of Bt's efficacy against certain pest populations in as few as two to four years."[21] Agribusiness apparently chooses not to see beyond the short-term profits available from selling crops engineered with Bt, despite their claims to be creating Bt products to improve the lives of farmers and consumers. Worse yet, this short-term exploitation of Bt may permanently eliminate its effectiveness for all growers, a move that has enormous implications for organic farmers, who will lose one of the most valuable tools they have against pests.

Pests can develop resistance because constant exposure to the Bt toxin allows them to develop a tolerance for it. This tolerance arises because the pesticide kills first the bugs most susceptible to it. The ones that survive – to mate and pass on their genes – are those most resistant to the poison. Thus, unless the pesticide kills every single bug, it functions over time as a mechanism to select for insects that are impervious to that pesticide. Resistant pests are sometimes referred to as "superbugs" because they can withstand high levels of pesticides.

The biotechnology companies who have commercialised genetically engineered Bt crops have been expecting pest resistance, and some have already developed and marketed a second generation of Bt crops and sprays. For example, AgrEvo's brand of Bt corn, called StarLink, uses a different Bt protein that "will retain [its] efficacy in a trans-genic crop system that is 'inundated solely with Cry1A toxins' (the usual Bt toxin), where resistance might develop."[22] At the same time that agribusiness corporations are assuaging the public, they are scrambling to create a solution to the growing problem of Bt resistance. Val Giddings, Vice President of the Food and Agricultural Division of the Biotechnology Industry Organization, calls concerns about pest resistance "bogus," claiming that technology will keep "well ahead of evolution of insect resistance for 100 years or more."[23] Given the fact that the entire international medical community has been unable to stay ahead of antibiotic-resistant organisms, it seems unlikely that the agribusiness corporations could stop pest resistance. Time and time again, the trial-and-error work of human researchers has proved no match for the incredible

Changing the Nature of Nature

evolutionary speed of trillions of natural viruses and organisms.

BT CROPS APPROVED FOR FIELD TESTING BY THE US DEPT. OF AGRICULTURE, 1987-1997[24]

Alfalfa	Peanut
Allegheny serviceberry	Poplar
Apple	Potato
Broccoli	Rapeseed (Canola)
Corn	Rice
Cotton	Spruce
Cranberry	Tobacco
Aubergine	Tomato
Grape	Walnut

GENETICALLY ENGINEERED FOOD AND HERBICIDES
The Monsanto corporation is playing both sides of the street, producing chemicals that kill plants and plants that resist being killed by chemicals. By creating food crops that are resistant to herbicides, you create a niche market for yourself and increase profits for the herbicide manufacturer. Of course, if you own the herbicide as well, then you create a closed-loop market for yourself – and that is exactly what Monsanto has done.

Monsanto sells farmers genetically engineered seeds, called Roundup Ready seeds, and often requires customers to sign a contract promising to use nothing but Roundup Ready seeds and the corresponding Roundup herbicide, also made by Monsanto. In addition, any farmer choosing Roundup Ready seed agrees to pay a "technology fee" per bag of seed, permit routine field inspection by Monsanto agents, and give up the right to save seeds for future planting. Such contracts are a blatant attempt to centralise agricultural control and ensure sales.[25]

Monsanto produces several Roundup Ready food crops, including corn and soybeans. Roundup is quite toxic; as Greenpeace notes, "If Roundup were sprayed directly on a normal soybean plant it would kill it."[26] Yet farmers who use Roundup can spray it again and again on their Roundup Ready soybean plants, which are resistant to the herbicide. Sold as a broad-spectrum weedkiller, Roundup is designed to kill virtually everything green in the field except the genetically engineered plant.

The herbicide not only affects the soil and water, but also affects habitat for wildlife, as it kills everything surrounding and among the crop plants. Without proper segregation, the soybean harvest from these "Roundup" crops end up in any number of foods that contain processed soy ingredients. Yet these genetically engineered food crops did not undergo independent, long-term safety tests to conclude that they are safe for human consumption prior to being put on the market.

Luke Lukoskie, owner of an organic soybean farm in Washington state, had this to say about Monsanto's Roundup Ready soybeans: "There are 10,000 natural varieties of soybeans – plenty to choose from – we simply don't need a genetic mutation of an already exceptional selection."[27] As we are all part of the natural world, the multiple doses of chemicals agribusiness companies dump on our food crops, land, and even our groundwater have profound consequences. Why, when alternatives exist to manage pests and weeds, do corporations continue to choose toxic methods of farming?

This technology has other pitfalls. Although the Environmental Defense Fund and Al Gore have portrayed glyphosate – the active ingredient in Roundup – as safer than other herbicides, its environmental consequences are certainly not trivial. Not only has glyphosate been documented as the third most common cause of illness among agricultural workers in California, but, according to the *Journal of Pesticide Reform*, it also harms beneficial fungi and nitrogen-fixing bacteria, both of which are essential for plant survival. The environmental effects are long-lasting: scientists have found herbicide residues in lettuce, carrots, and barley planted as long as a year after the soil was sprayed. In addition, there is also the threat of herbicide "drift," forcing neighbouring farmers to switch to Monsanto Roundup Ready seeds to protect their crops from the airborn herbicide – something Monsanto would surely not object to.[28]

THE TERMINATOR
What could agribusiness corporations think of next to alter our food supply? The answer: "Control of Plant Gene Expression," a benign-sounding name for a technique in which genetically altered plants are programmed to kill their own seeds. The Rural Advancement Foundation International (RAFI) dubbed this the "terminator" technology, since it eliminates a plant's natural ability to regenerate.[29]

More than 1.4 billion farmers around the globe rely on saved seed, but any farmers who use terminator seeds will be forced to buy new seeds for every planting, ending an age-old tradition of seed saving and creating a perpetual cycle of dependence on big seed companies. The terminator technology was created jointly by the United States Department of Agriculture (USDA) and Delta & Pine Land, a company Monsanto has acquired. Together they have been granted US patent #5,723,765, which is so broad it allows terminator technology to be used in the plants and seeds of all crops. The Rural Advancement Foundation reports that there are twenty-nine terminator patents, including two held by Monsanto, a dozen by Swiss pharmaceutical giant Novartis, and two by US public universities.[30]

Why would an agribusiness company develop a plant that kills its own seeds? The terminator appears to be in clear opposition to the natural cycle of plant life and plant regeneration, and it offers no benefit to farmers. But it does ensure corporate seed sales.

Outraged that public resources and money were used for the development of this technology that so blatantly serves private interests, over 7,000 people have written to the USDA expressing their opposition to the terminator technology. In a letter to the USDA, the Council for Responsible Genetics listed some of the negative impacts of terminator technology: "Sterile seed pollen can drift from field to field, carried by the wind or by insects, rendering nearby crops sterile. In addition, to commercialise these seeds they may need to be treated with antibiotics like tetracycline, a practice which could threaten the medical use of these antibiotics."[31]

The terminator continues to be opposed by thousands of citizens, scientists, and farmers. More recently, the Consultative Group on International Agricultural Research (CGIAR), an international plant-breeding network, rejected the terminator technology. They will not use the technology in their crop-breeding programs, and their pre-emptive decision has sent a clear message to those countries in which terminator patents are pending.

In May of 1999, New Hampshire became the first state in the United States to consider legislation banning the use of terminator technology.

Bill #291 was introduced by Representative Marie Rabideau, a gardener who learned about the terminator technology and felt compelled to introduce legislation establishing a committee to review the technology and advise state growers. Rep. Rabideau called for a precautionary ban on the terminator technology in New

Hampshire and a system of penalties for corporations and farms that use terminator seeds. The bill also required the state commissioner of agriculture to complete an annual report on recent developments in the field of genetically engineered plant life. One of the experts who testified in favour of the bill was Martha Crouch, a professor of biology at Indiana State University.[32] Professor Crouch ended her testimony by telling the committee that if they banned the terminator technology, the New Hampshire state motto, "Live Free or Die," could ring true for seeds too.

Monsanto's pledge in late 1999 not to use the terminator gives some hope that the holders of the other twenty-seven terminator patents will follow suit.[33]

MILK ADDITIVE: rBGH

> I was told that I was slowing down the approval process. It used to be that we had a review process at the FDA. Now we have an approval process. I don't think the FDA is doing good honest reviews anymore. They've become an extension of the drug industry.[34]
>
> **Former FDA employee** Dr. Richard Burroughs, fired after ordering toxicology and immunology tests on rBGH

Recombinant Bovine Growth Hormone (rBGH), or Bovine Somatotropin (rBST), is a genetically engineered hormone that tricks a cow's body into producing more milk than it otherwise would. Milk from cows treated with rBGH has been known to be contaminated with pus from udder infections, with antibiotics administered to stem those infections, and with high levels of insulin-like growth factor (IGF-1), which has been linked to human breast and gastrointestinal cancers.[35]

IGF-1, a hormone protein that is present in humans, helps cells divide. Of course, high levels of cell division and growth are undesirable: as the *Lancet*, the pre-eminent British medical journal, reported in May 1998, a high level of IGF-1 is a serious risk factor in the development of breast cancer. The same issue of the *Lancet* implicated IGF-1 in colon and prostate cancer as well. The presence of elevated levels of IGF-1 in milk from rBGH-injected cows suggests that serious caution should have been taken before permitting release of this substance into the food supply. Some scientists believe the levels of IGF-1 are much higher in milk from rBGH cows than in regular milk, and it is well-known that IGF-1 is still present in milk after pasteurisation. In fact, some studies have shown that the level of IGF-1 can increase as much as 70 percent during milk pasteurisation, and the meat of cows treated with rBGH has also been found to contain unusually high amounts of IGF-1.

rBGH, which was developed by the Monsanto corporation, has serious health implications for cows as well. Problems include "cystic ovaries, uterine disorders, decrease in gestation length and birth weight of calves, increased twinning rates and retained placenta."[36] In 1999 the Canadian government banned the use of rBGH, basing their decision on animal welfare issues such as the increased occurrence of udder infections and decreased life span of cows treated with rBGH.

Monsanto's application was controversial from the start over eight years ago, when consumers and farmers alike began voicing their opposition to the use of rBGH. Claims of corruption began to circulate in Canada, including an allegation that Monsanto offered HealthCanada scientists research money in exchange for their approval of rBGH. The Canadian Broadcasting Corporation (CBC) aired a story in which Monsanto officials allegedly offered HealthCanada officials two million dollars to approve rBGH.[37] The issue was brought to a head in late 1998 when six scientists at HealthCanada produced a report aptly titled the "Gaps Report," which chronicled the history of Monsanto's application within HealthCanada and cited the many missing documents, or gaps, in the normal approval process, including faulty experiments and incomplete data.[38]

The Gaps Report states that Monsanto did not subject rBGH to any of the normally required long-term toxicology experimentation and tests for human safety.[39] The report also suggests that Canadian regulatory officials made an exception for Monsanto by not requiring that it submit the appropriate tests to make sure that rBGH was safe for human consumption. The report has called into question not only the lax and incomplete application process, but the conduct of senior officials at both the US government, which approved rBGH, and the Canadian government, which eventually did not, all of whom seemed to show the application unusual favour. The Gaps Report ends with the following statement: "Duly arising from this particular issue certain senior officials of both these agencies have allegedly been asked to be investigated for employing unauthorised influence against subordinate staff and a personal conflict of interest."[40]

The US government approved the use of rBGH in 1993, to the chagrin and outrage of farmers, activists, and consumers. Not surprisingly, the Canadian Gaps Report questions the FDA's approval process, noting that "The United States is the only developed country permitting the use of rBST, of which there are four manufacturers. There are reports on file that Monsanto pursued aggressive marketing tactics, compensated farmers whose veterinary bills escalated due to increased side effects associated with the use of rBST and covered up negative trial results."[41] rBGH/rBST manufacturers include Monsanto, Elanco/Eli Lilly, American Cyanamid, and Coopers Agropharm.

Opposition to rBGH has escalated in the United States. Dairy manufacturers have found a receptive market for rBGH-free dairy products, and more companies are taking it upon themselves to label their products as such. *The New York Times* reported that organic milk sales have skyrocketed in the United States, from $16 million in 1996 to nearly $31 million in 1997.[42] US consumers have good reason to be wary of rBGH milk, as many of the tests for human health and safety that were conducted on laboratory rats led to the development of cysts on the thyroid and prostate when rBGH damaged the rats' immune systems.[43]

USE OF ALLERGENS

In 1996 researchers at Pioneer Hi-Bred were working on a genetically engineered soybean that contained a protein from Brazil nuts (methionine). The resultant soybeans, genetic engineers hoped, would have greater nutritional value. Since it is widely known that Brazil nuts can cause allergic reactions in people, this group of scientists decided to test their genetically engineered soybeans for an allergic response in humans. Allergic reactions occurred as reported later that same year in the *New England Journal of Medicine*.[44] Simply put, if you are allergic to Brazil nuts and are eating soybeans genetically engineered with Brazil nuts, you will have an allergic reaction. For some people who are allergic to Brazil nuts, eating this type of genetically engineered soybean could be lethal.

Dr. Rebecca Goldberg, a senior scientist at the Environmental Defense Fund, says, "Since genetic engineers mix genes from a wide array of species, other genetically engineered foods may cause similar problems. People who are allergic to one type of food may suddenly find they are allergic to many more."[45] Interestingly, this allergy test was not required by any of the regulatory agencies of the US government; it was a precautionary test conducted at the discretion of the scientists involved. When Pioneer learned of the results of the allergen test, it withdrew the product.

While genetic engineers seem to have stopped experimenting with known allergens, they do use novel plants, viruses, and bacteria that humans do not generally consume. For example, petunia, genes of which are used in Monsanto's Roundup Ready soybeans, has not been a part of most Americans' diet. Do you know if you are allergic to petunias? The effects of eating certain species are simply not known. In essence, we are guinea pigs in this experiment, as our reactions to GM food will help classify these novel food products as allergens.

Plant species that are not "known" allergens should not be assumed safe without testing for allergenicity. Because many transgenic crops end up as animal feed, we should also take into consideration the health and safety of the animals that may consume transgenic foods. Furthermore, we should take into account the health and safety of those people who may in turn consume the animals that ate genetically engineered food.

GENE JUMPING

> Gene flow refers to the potential spread of newly inserted genes into related wild species. For instance, if insect resistance is engineered into a sunflower crop, those new sunflowers could reproduce with wild sunflowers to create sunflowers that are different from the original wild sunflowers. Eventually, we might completely lose the original sunflower gene pool."[46]
>
> Environmental Nutrition

Genes do not necessarily remain in organisms, or stay where they are put. They can "flow" through natural cross-pollenisation to related organisms, or they can move between unrelated organisms using viruses and some bacteria as vectors. They can travel outside the original organism to infect another. This process is known as horizontal gene transfer,[47] or species "jumping," and it is another inherent danger of genetically altered foods. When genes from the genetically altered crop "jump" to neighbouring crops or crop relatives such as weeds, they can create what some people call "superweeds." These superweeds have some of the characteristics of the genetically engineered crop, such as herbicide resistance, pesticide resistance, or viruses. Superweeds can then pass their genes on to succeeding generations, and the cycle continues, into what can be called "bio-pollution." Genes naturally flow among related plants – it is a common and relatively harmless process in nature. What separates genetically engineered crops from others is that they contain completely foreign genes. Once these genes escape into wild populations, there is no putting the genie back in the bottle.

ANTIBIOTIC RESISTANCE
Antibiotic "marker" genes, such as the one used in the Flavr Savr tomato that conferred resistance to the antibiotics kanamycin and neomycin, are used frequently in genetic engineering. Novartis Corporation's Bt maize, as well as Monsanto/Delta & Pine Land's terminator technology, specify the use of antibiotics. Foods that contain antibiotic-resistant genes could pass this resistance on to disease-causing bacteria by passing their resistant genes to the naturally occurring bacteria they encounter in the human digestive system, which in turn share the genes with human pathogenic bacteria – a dangerous situation, since our greatest weapons against bacterial infections are antibiotics. Serious health problems could ensue as traditional antibiotics would no longer be effective against bacterial infections. For example, the genetically engineered bacterium *B. subtilis* contains a kanamycin resistance gene. Kanamycin resistance also can result in cross-resistance to the antibiotics amikacin and tobramycin, both of which are extremely important to fighting infection.[48]

Since a wide variety of cellular mechanisms are targeted against the transfer of foreign DNA between plants, scientists had to overcome natural species barriers. To do this, they created artificial gene-transfer vectors that are hybrids of viral DNA and bacterial DNA. These artificial vectors are aggressive in transferring their foreign genes into the desired organism – be it bacterium or maize. As mentioned previously, antibiotic-resistance marker genes are included in the transfer vector because conferred antibiotic resistance is a simple indicator of successful transformation: only the organisms that survive when treated with the antibiotic contain the gene of interest.

The aggressive nature of these gene-transfer vectors is what makes them so dangerous. There is no guarantee that these genes will stay put when introduced into a new organism, because they are designed *not* to stay put. Furthermore, "genes carried by vectors can survive indefinitely in the environment, within thriving or dormant bacteria, or as naked DNA adsorbed to solid particles, where they are efficiently taken up by other microbes."[49] Thus, the potential for an uncontrollable spread of antibiotic resistance through genetic engineering is a sobering possibility.

In 1998 researchers from Indiana University reported in the Proceedings of the National Academy of Sciences USA that a genetic parasite of yeast had jumped into many unrelated higher plant species. Such horizontal gene transfers have already been reported in bacteria – the culprits behind the rapid resurgence of antibiotic-resistant infections.[50]

With such risks, the release of Novartis's Bt maize in Europe has sparked a fierce debate over the use of antibiotic-resistant genes in agriculture. In this case, a gene was inserted into maize that confers resistance to the antibiotic ampicillin. Ampicillin is a member of the penicillin family of antibiotics – the most clinically used antibiotics. The artificial ampicillin-resistance gene introduced into Novartis's Bt maize enables the breakdown of a number of penicillins, including ampicillin, penicillin G, amoxicillin, penethicillin, carbenicillin, methicillin, and cloxacillin – drugs heavily relied upon to fight serious infections such as pneumonia, bronchitis, and diphtheria. Widespread resistance to these antibiotics would have monumental consequences. Why take the risk of increased antibiotic resistance when alternative genetic markers are available that have no effect on antibiotic activity? The release of such resistance markers is not only unnecessary but extremely irresponsible.

Nearly 60 percent of scientists polled by the International Society of Chemotherapy believed corn genetically engineered for antibiotic resistance to be an unacceptable risk for consumers. In addition, 34 percent said that more risk assessment was necessary *before* food crops treated with antibiotics are brought to market. Only 2 percent of those polled deemed the maize safe for consumption.[51] In July 1999 the Swiss government banned genetically modified maize that was undergoing field trials in Britain on the grounds that further study to establish the safety of its antibiotic-resistant genes and other genetic modifications was needed.[52]

Another genetically engineered crop that carries the threat of widespread antibiotic resistance is Monsanto's terminator crop, designed to kill its own seeds. Dr. Martha Crouch, a molecular biologist at Indiana University, explains that, in order to activate the terminator genes, seed companies will have to soak the seeds in the antibiotic tetracycline prior to their sale to farmers. According to Dr. Crouch, "Large-scale agricultural uses of antibiotics are already seen as a threat to their medical uses."[53] Tetracycline is a relatively non-toxic drug with a wide variety of clinical applications that include the treatment of rickettsial infections, chlamydia, pneumonia, genitourinary tract infections, and malaria. Widespread tetracycline resistance would be a severe loss, perhaps requiring the hazardous use of stronger, more toxic drugs to treat such infections.

The use of antibiotics in food crops is reminiscent of a case in East Germany in 1982 where the antibiotic streptothricin was given to pigs on a large scale. Within a year, genes encoding resistance to streptothricin were discovered in pig-gut bacteria. By 1984 this resistance had reached the gut bacteria of farm workers and finally, in 1985, it reached pathogenic bacteria in the general public. This phenomenal spread of resistance forced the withdrawal of the antibiotic in 1990.[54]

OWNERSHIP OF LIFE
> A veil of secrecy made heavy with money has fallen over grain research and biogenetics. And it's a phenomenon that reaches from the largest industrial nations to the smallest rural countries.[55]
> Hannelore Sudermann, Spokesman Review

Orville Vogel, formerly a plant scientist at Washington State University, was renowned for his development of a successful high-yielding wheat variety. This new wheat variety, which he developed in the 1950s, was a technology made freely available to farmers. It doubled grain yields, earning Northwest farmers tens of millions of dollars.[56] Those days of open sharing are history. Now many scientists and plant breeders are more apt to patent their new plant variety, keeping it secret even from colleagues and fellow scientists until the patent has been granted.

Today genetic engineering of food crops is mainly oriented toward profit, and patents are an integral part of insuring those high returns. You may be wondering how a plant – a life form – can be patented? How did nature's wonders become the private property of multinational corporations? You can patent a mousetrap, a mechanism, or handy new tool, but a plant? It may be hard to believe, but in a little known US Supreme Court decision in 1980 patents were allowed on living organisms. The case, Diamond v. Chakrabarty,[57] which granted ownership rights over a microorganism, has been used as a rationale for the continual expansion of patent rights – to microorganisms, plants, animals, and even human genes.

Calgene patented its Flavr Savr tomato in 1990. Since then, US patents have protected virtually all genetically engineered food crops. There are patents on the naturally occurring soil bacterium Bt and its use in genetically engineered crops. By mapping out Bt's molecular properties, agribusiness corporations can claim to have "discovered" this wild soil bacterium, at least as far as patents on the organism and its use are concerned.

In the mid-1990s, citizens groups and non-governmental organisations began to work together to oppose the ownership of life. In 1996, a coalition issued a statement and petition saying:

> The plants, animals and microorganisms comprising life on earth are part of the natural world into which we are all born. The conversion of these species, their molecules or parts into corporate property through patent monopolies is counter to the interests of the peoples of this country and of the world.
>
> No individual, institution or corporation should be able to claim ownership over species or varieties of living organisms. Nor should they be able to hold patents on organs, cells, genes or proteins, whether naturally occurring, genetically altered or otherwise modified.
>
> As part of a world movement to protect our common living heritage, we call upon the Congress of the United States to enact legislation to exclude living organisms and their component parts from the patent system.[58]

Patents have an enormous impact on the development of genetically engineered food. They affect the price and availability of seed, and they force patent holders to play an "enforcement" role, protecting their exclusive patent rights with lawsuits and legal agreements. When a single company such as Empresas La Moderna owns 25 percent of the world seed market, we face unprecedented domination of agriculture by multinational corporations. A partnership of farmers, concerned government officials, scientists, and those of us who eat food – in other words, all of us – can take back our seeds and restore the food security, health, and safety that each of us needs and deserves.

3
YOU ARE WHAT YOU EAT

Throughout history monarchs have employed food tasters. This rather high-risk line of work was invented not for gastronomic reasons, but out of a recognition that when we eat food we are placing a great deal of trust in whomever provides that meal. In societies where people grow their own food one has a pretty good idea of the origins of the food, what was sprayed on the crop as it grew, and how it was cooked. When food is produced locally, just keep the peace with the farmer and the chef and you can eat your dinner with no worries.

In nearly all societies nowadays, even monarchies, most people no longer grow their own food. We eat our meals each day with the assumption that what is sold is safe, both because we choose to trust the farmers and food sellers and because we have some degree of faith in government regulators and food inspectors. If this system of trust breaks down, people would understandably be frightened.

Recently, one of the authors was preparing a huge pot of vegetarian chilli for a group of friends. Opening a large can of red kidney beans from a local supermarket, he was surprised to find the can full of peeled, white potatoes. Opening another can of beans from the same store, he found the same unexpected contents. Every person at the chilli party had the same reaction: "Don't ever shop at that place again." Even though this one error in labelling some cans of beans produced no illness and was presumably an isolated incident, every person seeing the mislabelled cans had the same extreme reaction: their assumptions that food is carefully monitored all the way into their supermarket trolleys were temporarily destroyed.

The GM food industry knows of the tendency of people to have a short fuse when it comes to food safety issues. A Monsanto spokesman told one of the authors that the firestorm of protest over GM foods in Great Britain in 1998 and 1999 was probably strongly propelled by the recent scare over mad-cow disease in that country. This statement recognises that even though mad cow has little to do with genetically engineered food – any more than a mislabelled can of red beans has anything to do with a store's fish or lettuce – when trust in food purity and safety is shattered, people become *very* conservative.

And so they should. In the United States, up to 80 million people are estimated to become sick from food – caused illness each year. Nine thousand of them die.[1] Statistics from other countries, when they are available, are comparable. The real incident of illness is probably much higher: many of us have come home from a barbecue featuring sun-baked potato salad, or tried out a new restaurant that looked a bit seedy, and then attributed that night's sickness to the food we just ate. These relatively minor instances of food-related illness never get reported to the collectors of statistics. We just take some pinkish medicine and wait for the bad time to pass.

Other food-connected illnesses are less benign. So far, more than twenty deaths are attributed to mad-cow disease in Great Britain, and the number of people ultimately affected by this slow-moving disease won't be known for some time.[2] A whole class of deadly illnesses that might be related to some food is cancer, possibly arising from the residue of added chemicals on or in the food. Because studies linking slow-onset diseases that have complex causes are still underway, contradictory, or not readily available, some people choose to ignore the possibility of risk until better information is available. Others decide on a more cautious approach and go to the extra trouble and expense of purchasing foods labelled "organic" to avoid consuming those chemicals.

Many food-related illnesses lie somewhere in between a minor bellyache that goes away in a day and tragic death from chemical-triggered cancers. So-called "Chinese restaurant syndrome" comes from ingesting too much monosodium glutamate (MSG), a flavour enhancer used in many prepared and Asian foods. Because people have significantly varying sensitivities to various foods, some restaurants take the precaution of not adding MSG to any of their food and post signs telling their customers of this choice. As we'll see in this chapter, the great variability of individual reactions to ingested substances – some people show MSG reactions from relatively small amounts of the additive while others have no discernable symptoms – should lead us to an overall abundance of caution when we are considering food safety policy in general.

In this chapter, we are going to look at the human health issues associated with genetically engineered food. Is it safe to eat GM food? Let's look at the information available, and then see what actions a person might take.

ALLERGIES

As we discussed in chapter 1, each gene contributes a single protein to the genetic "soup" that comprises a living organism. Proteins are crucial substances that play many roles in human physiology. One clear association with proteins involves allergies. When a person exhibits an allergic reaction, what her body is reacting to is a protein, most often a "foreign" or introduced protein.

This leads us to a serious issue that arises in connection

to genetically engineered foods. If allergies are associated with introduced proteins, and if GM food is by definition characterised by introduced genes that produce proteins, then we have a situation in which caution about allergies is justified. As we saw in chapter 2, allergies have already been proved to pass from one type of food into another via gene transfers. The fear of introduced genes triggering dangerous and even fatal allergic reactions is based on sound science.

Allergies are common in people, ranging from extreme reactions to exotic fish to a mild sensitivity to airborne pollen in the springtime. The amount of a given protein that might trigger an allergic reaction is highly variable. Some people are allergic to common foods like wheat or eggs, while others are able to eat them with no ill effects. Most of us eat peanuts with pleasure, while a few people can find themselves fighting for their lives when they consume even a miniscule quantity of peanut protein. Because of this, most allergy specialists do not advise patients with food allergies to cut down on the food that they are sensitive to. Instead, they sternly admonish their patients to avoid *any* ingestion of that food in even the tiniest, most insignificant-seeming quantity.

The great solace – and safety – for people with food allergies is the labelling of ingredients. In the United States and in many other countries, food producers are obligated to list the ingredients of any prepared or processed food. People with allergies avoid frantic trips to the emergency room by learning to read package labels carefully. Food manufacturers and distributors avoid costly liability by this same disclosure of contents. Our laws tend to say that if an ingredient is revealed, that constitutes fair warning. So if people suffer ill effects from eating something that was properly labelled, they cannot sue the food producer.

This is not so with genetically engineered food. Even in countries that require GM food labelling, the labels will most often just say that genetically modified substances are in the food container. Because there is no requirement to say which gene has been inserted, people must avoid all GM food if they have allergic sensitivities and want to be totally prudent.

It is important to note that, unlike mad-cow disease, there have been no documented cases of deaths due to GM food-caused allergic reactions.

However, because an autopsy for a death from allergic shock does not normally test for the presence of genetically engineered food, there is no reliable way of gathering data on GM food allergic reactions.

Genetically modified foods available around the world do not present enhancements to the buyer and consumer of those foods. The foods do not taste better, provide more nutrition, cost less, or look nicer. Why, then, would a person with a food allergy run the risk, however large or small it might be, of a life-threatening reaction when safe alternatives are available? We just need to make sure that those alternatives remain available.

NUTRITION

The assumption that many of us would make is that GM food is nutritionally equivalent to non-modified food. In 1999 the California-based Center for Ethics and Toxics (CETOS) set out to see if this was the case. The people at the centre noticed that the research submitted to, and accepted by, the US government to demonstrate the safety of Monsanto's genetically engineered soybeans had been conducted by Monsanto's own scientists. A conflict of interest doesn't necessarily mean that people with the conflict are dishonest, only that, as in this case, their associations automatically put the objectivity of their work into question. The CETOS staff, Britt Bailey and toxicologist Marc Lappé, observed that the soybeans Monsanto tested were not an accurate representation of the soy that appears in stores as food because they were not treated with the herbicide Roundup, as they would be in real life. So Bailey and Lappé hired a reputable testing firm to conduct tests that would accurately compare Roundup Ready soybeans, treated with Roundup, to conventional soybeans that were identical to the Roundup Ready ones except for the missing engineered gene. The tests were also carefully designed to produce results that reflected real-world conditions. This sort of objectively designed science is what we need to be able to make good decisions about what we buy and eat.

The study was published in a peer-reviewed scientific journal in 1999. The process of peer review is important in science. It means that independent scientists looked at the CETOS study and found it to be based on sound and acceptable methods of scientific investigation. In their study, CETOS found that there was a 12 to 14 percent decline in types of plant-based estrogens called phytoestrogens. Phytoestrogens are associated with protection against heart disease, osteoporosis (bone loss), and breast cancer. A drop in phytoestrogens of 12 to 14 percent is a significant nutritional difference.

The CETOS study was attacked by the American Soybean Association, whose attack was in turn answered by CETOS. Monsanto also conducted new studies that did not show the same changes in phytoestrogens. The new Monsanto study is difficult to compare with the CETOS study, however, because Monsanto inexplicably used a different, older method in some of its research. Meanwhile, CETOS stands by its study, which the researchers point out at the very least raises some important questions about nutritional variances in this particular food.

While scientists sling studies and journal articles at each other, what's a food shopper to do? We can't all be expected to become experts on obscure scientific methods or substances we never knew existed, like phytoestrogens. The government, which chose to accept the original

Monsanto-paid tests as the basis for approval of this food, has been of no use in helping us to make prudent decisions.

Further, we have no way of knowing if differences in plant hormones in soybeans mean much for human health, and more important, if the CETOS study was a fluke or will turn out to be typical if CETOS, a non-profit organisation, can garner sufficient support to conduct more scientific experiments.

What can we conclude about nutrition and GM foods from this example?

Soy products appear in many processed and prepared foods. No one knows just how many, but the words soybean oil, soy flour, isolated soy protein, textured vegetable protein, functional or non-functional soy protein concentrates, and textured soy protein concentrates on the label are good tip offs. As much as 60 percent of all prepared food in a typical US supermarket contains genetically engineered ingredients.[3] Further, many people who do not eat meat rely heavily on soy-based food for important nutritional components of their diet, including proteins, some fats – and phytoestrogens.

As with allergies and GM food, we are left with more questions than answers. Is genetically modified food more nutritious? There is no evidence for a claim of this type. Is genetically modified food less nutritious? We do not know for certain, yet. Is genetically modified food perhaps more variable in its nutritional value? We have at least one reputable study that suggests yes.

The conclusions we can draw from what is known, and not known, seem to be fairly straightforward. First, there is clearly a great need for many further studies of possible nutritional changes in genetically engineered food, based on the CETOS study. Second, there should be clear, unequivocal labelling of GM food so that people can make their own decisions about their nutrition.

PESTICIDES AND HERBICIDES

A potential problem arising from herbicide resistant GM crops that is largely being ignored is what is the fate of these chemicals within the plant? Are they stable? If they are degraded, what are the products that are produced? And what health risks do they pose?[4]

Michael Antoniou

It should be no surprise to us that in discussing genetically modified food we need to pay attention to chemicals that are designed to kill plants or animals. Many of the significant GM food crops are engineered to either tolerate higher than usual amounts of herbicides or to contain pesticides inside each cell. As we saw in the first two chapters, many of the purveyors of GM food are companies that market agricultural chemicals. Engineering plants to require what a company already sells makes business sense to these corporations.

The pesticide most in question is *Bacillus thuringiensis*, Bt. This bacterium was isolated one hundred years ago, although it did not become commercially available in the United States until 1958.[5] While this bacterium is related to a common bacterium that causes food poisoning and is also a close relative of the organism that causes anthrax, Bt itself is considered relatively safe, especially when compared with synthetic bug-killing chemicals.

Yet Bt *is* a poison. In its purified form it can be extremely toxic to mammals, including humans, and even in its more usual, non-purified state there are numerous reports of poisonings and various negative health effects. The chemical is thought to be particularly hazardous to people with compromised immune systems.[6] Yet because the EPA has already established that Bt (as a spray *on* crops) is safe, it assumes the toxin is safe to eat *in* crops and does not require testing for human health effects.

When Bt is used by organic growers, it is sprayed on

plants. It breaks down rapidly in the environment after killing the target bugs. While there may be health risks to the person applying the pesticide, there is no clear evidence of health problems resulting from people eating food that has been sprayed properly with Bt.

When Bt is engineered into a plant, it may remain present in the plant, and the resulting food, much longer than is the case with conventionally used Bt. There is even evidence that Bt engineered into plants remains after the harvest, so that plant leaves that drop to the ground or plant residue that is ploughed under have an effect on the living organisms in the soil.

What we do not have is a series of clear, independent studies on the long- or short-term health effects of eating food containing the pesticide Bt. According to CETOS, from 1987 through 1998 24 percent of genetically engineered crops released into the environment contained insect-resistant genes. According to this same source, Bt crops are grown in the United States, Brazil, Argentina, China, India, Australia, Canada, South Africa, and Japan. Yet we do not have information in hand to establish the safety of this pesticide for human health. The EPA does not test the plant with Bt in it, it only tests the bacteria in isolation. Essentially, the EPA is not testing the product that humans will be consuming. The Bt toxin produced by the plant and the toxin produced by the bacteria could be different. Until both are properly tested for human health effects, no one will know the effect of eating Bt food crops.

In a case that gained some notoriety in early 1999, a scientist at the Rowett Research Institute named Arpad Pusztai tested genetically engineered potatoes on rats. After only ten days the animals suffered substantial health effects, including weakened immune systems and changes in the development of their hearts, livers, kidneys, and brains.

When Dr. Pusztai went public with his findings, he was

summarily fired and a commission was convened by his former employers to investigate his work. The commission found Dr. Pusztai's work deficient, yet another panel of twenty independent scientists confirmed both his data and his findings. More recent research by another UK-based scientist showing enlarged stomach walls in rats fed genetically engineered potatoes seems to support Dr. Pusztai, who has stated publicly that he will not eat GM food.

Aside from Bt crops, the other major genetically engineered plant chemical involves herbicide-tolerant plants, primarily the Roundup Ready series of plants from Monsanto. These genetically modified plants include corn and soy as well as oil-producing canola (rape seed) and cotton. As we saw, the plants are engineered to withstand the plant-killing effects of the chemical glyphosate, the main ingredient in Roundup. Monsanto claims that this herbicide tolerance means that farmers can spray the plant-killing poison on their fields more precisely and thus use less, but there are serious concerns about how much herbicide is actually being sprayed.

What about the health effects of herbicide-tolerant crops? Scientists have already linked the herbicides containing glyphosate to cancer. Non-Hodgkins lymphoma, which is one of the fastest-rising cancers in the Western world, increasing 73 percent since 1973, has been connected to exposure to glyphosate and MCPA, another common herbicide.[7]

Since even proximity to such chemicals has been linked to cancer, what are the health risks of eating crops sprayed with glyphosate or genetically engineered with Roundup resistance? While the maker of Roundup insists that when used properly the herbicide is safe, independent studies raise a long list of questions about the long- and short-term health effects of human ingestion of glyphosate. Some European studies are downright alarming.[8]

Bottom line: genetically engineered food hasn't been proved safe. Since wholesome alternatives exist, why not suspend production of GM food until it is shown to be wholesome?

ANTIBIOTIC RESISTANCE
As we have seen, some crops are genetically engineered with antibiotic-resistance markers, which enable scientists to test for and track genetically engineered characteristics. Perhaps inadvertently, these markers have also been used to find farmers who are allegedly growing genetically modified crops without having paid the license fees.

Other substances can be used as markers, but antibiotics are convenient.

The problem is that the use of antibiotic-resistant marker genes in food means that people are eating antibiotic-resistant genes, and perhaps the healthful bacteria in their bodies are taking on resistance to antibiotics along the way. Each year, approximately five billion dollars are spent on the treatment of antibiotic-resistant infections in the United States. The threat of enhanced antibiotic resistance among disease-causing pathogens is more real than ever before as bacteria develop ways to elude even the strongest drugs.

Antibiotics are an effective weapon against disease. However, their efficacy has rapidly fallen with time. For example, in 1952 all cases of staphylococcus infection responded to penicillin treatment. By 1980, only 10 percent of all treatment cases were successful.[9] Overuse of prescribed antibiotics is one of the main culprits behind this increased resistance, as pathogens "learn" to develop chemical resistance mechanisms in response to continuous exposure. As harmful bacteria become more resistant to traditional antibiotics, the need to restrict the unnecessary use of antibiotics is crucial. According to the Genome Therapeutics Corporation, "nearly 9 million people in the

US are affected by drug resistant bacterial infections each year and [these infections] are the cause of death for approximately 60,000 of these individuals."[10] Antibiotic resistance among old strains of disease is raising both the cost of treatment and number of deaths each year.

While overuse of antibiotics is a serious problem, another growing public health threat is the widespread use of antibiotics in farming. For decades, antibiotics have been administered to livestock to improve the quality and shelf life of meat, eggs, and dairy products. However, this use has not been without harmful consequences. For example, in the 1970s resistant salmonella strains were found in the meat and eggs of chickens treated with antibiotics. Less than a decade later, antibiotic-resistant salmonella that attacked humans had appeared on the scene.[11]

In the face of the misuse of antibiotics in medicine and agriculture, scientists have tried to learn about the origin of antibiotic-resistant genes. There is still a great deal that is not entirely understood. Many genes appear to arise spontaneously, some from mutations in pre-existing genes.

Regardless of their origins, antibiotic-resistant genes have displayed a remarkable capacity for spreading between organisms by jumping species barriers. Particularly significant is the fact that bacteria share genes all the time. Robert Havenaar, at the TNO Nutrition and Food Research Institute in the Netherlands, engineered bacteria to contain antibiotic-resistant genes. He found DNA from the bacteria to have a half-life of six minutes in the colon – ample time to get picked up by other bacteria.[12]

If antibiotic-resistant genes can indeed jump from plants to pathogenic bacteria, as is suggested but not proved by this research, then we are left with the possibility of human beings finding that their ability to be cured of infections by certain antibiotics could decline or even disappear as people eat food containing antibiotic-resistant genes.

BOVINE GROWTH HORMONE

Sold in the United States under the Monsanto brand name Posilac, rBGH is injected into cows to increase milk production. Few people would argue that the drug does increase milk production, although in a country that periodically gives away dairy products to deal with the milk surplus, it is difficult to understand why we need even more. Aside from well-documented health problems for the cows, including increases in udder infections, there are a series of health issues for humans.

As early as 1995, at an NIH conference, the following adverse effects of rBGH were identified:

1. Strong role in breast cancer
2. Special risk of colon cancer due to local effects of rBGH on the GI tract
3. May play a role in osteosarcoma, the most common bone tumour in children, usually occurring during the adolescent growth spurt
4. Implicated in lung cancer
5. Possesses angiogenic properties – important to tumours, some of which secrete their own growth factors to promote angiogenesis
6. Lastly, the 1995 NIH conference recommended that the acute and chronic effects of IGF-1 be determined in the upper GI tract.[13]

Americans have been drinking milk from cows treated with rBGH for several years now. When the hormone was approved by the US government, the approval was based on studies of rats fed rBGH that showed no toxicological changes. Had there been any such changes, further human studies would have been mandated.

In the well-publicised 1998 Canadian Gaps Report discussed in the last chapter, we learn that in fact a large proportion of the rBGH-fed rats, between 20 and 30 percent, showed distinct immunological changes, while some male rats showed the formation of cysts of the thyroid and infiltration of the prostate. These are warning signs for possible immune system effects – and possible carcinogenic effects as well.

The Center for Food Safety and more than two dozen other organisations filed a petition in December of 1998 to reverse FDA approval of rBGH/rBST: "We're going to go to the courts and say – you were lied to," said Andrew Kimbrell of the Center for Food Safety. "Essentially it was fraud by the agency and fraud by Monsanto in telling the court that there were no human health effects possible from consuming these products made with rBGH-treated milk. We now know that not to be true." The Canadian Gaps Report, the banning of rBGH in many countries around the globe, and the findings of a number of studies in the United States and in Europe all point to real, concrete health concerns about bovine growth hormone.

Estimates are that 15 to 30 percent of the milk supply of the United States comes from rBGH-injected cows. Since rudimentary labelling of rBGH milk exists in some communities, including direct labelling as well as the labelling of some milk as "organic," people can avoid feeding their families dairy products containing genetically engineered growth hormones. Where such labelling or alternatives do not exist, there is little choice for people other than turning to chapter 9 in this book and becoming active in nationwide efforts to provide people with the option to consume only the food that they feel is safe for their families.

IS GENETICALLY ENGINEERED FOOD SAFE?
In this chapter we have examined a number of different possible health issues with genetically modified foods. In some instances, such as phytoestrogen decline in genetically engineered soy, or a variety of health questions arising from animal studies of bovine growth hormone, there are ample reasons for people to decide to avoid

genetically modified food. In other instances, such as the health effects of ingesting herbicide-tolerant engineered food, there just isn't enough good science yet to be sure.

If the FDA does not require labels, and safety testing is the exception rather than the rule, just what is the US government doing to protect the public? The Hoover Institute's Henry Miller, a fan of genetic engineering, writes, "The FDA does not routinely subject foods from new plant varieties to pre-market review or to extensive scientific safety tests."[14] Later he notes that the FDA only follows "the development of foods made with new biotechnology via non-compulsory informal consultation procedures."

The conclusion to all of this is clear. There is no genetically engineered food product on the market now – not one – that is necessary. Each product, which may confer financial benefit to its producers, can be shown to have an alternative that from the consumer's point of view is at least equivalent if not superior. If GM foods do not provide a benefit to consumers, and may be shown to have health hazards now or in the future, why take any risk with your health or your family's health?

Since safety has not been demonstrated and our health is precious, avoid eating all GM food. And read on, to learn more.

4
YOUR RIGHT TO KNOW

> I personally have no wish to eat anything produced by genetic modification, nor do I knowingly offer this sort of produce to my family or guests. There is increasing evidence that a great many people feel the same way.[1]
>
> Charles, Prince of Wales

Ranchers who raise cattle have to follow rules, such as limits on how close to the time of slaughter the animals can be given certain drugs, to avoid tainting the hamburgers with the residue of the drug. The slaughterhouses are monitored daily by federal inspectors. The fast-food chains that sell the hamburgers are licensed and inspected by local health departments, and their employees meet local and state health-department regulations. Yet in the face of all these rules and regulations, children still get sick from eating tainted hamburgers. In Great Britain, all the inspection in the world couldn't find a disease no one knew to look for – so-called mad-cow disease.

Hearing stories like these, people who want to feel secure with their food choices often come to rely on the one person they feel will not make sloppy mistakes or cut corners on health and safety: themselves.

But in order to make good decisions we need

information. Most food comes from distant places and is concealed within packaging. So we rely largely on what we are told about the food by labels on the packaging.

Because genetically engineered food is by definition changed in ways that are hidden, and because testing for the presence of inserted genes is hardly practical for the average shopper in a supermarket, we are at the mercy of the labels if we want to make choices about eating GM food.

While makers of genetically modified food resist labelling of such food, claiming that the information will just provoke incorrect decisions based on ignorance or emotion, most people want labelling of GM food, as we will see.

Labels of other ingredients, qualities, quantities, or features don't seem to be a problem. A bag of candy purchased in an American market is almost covered with a wealth of label information. The package face tells us how many lollipop candies are in the bag, how much each weighs, and how much the bag weighs in total, both in ounces and grams. There is a kosher symbol next to the word for kosher, *parve*. Then there is a list of what is *not* in the candy, interesting to note since GM food manufacturers are especially resistant to labelling food GMO-free. The label says:

- No sugar
- No Saccharin
- No NutraSweet
- No artificial flavour
- No artificial colour
- No salt
- No fat
- No cholesterol
- No caffeine added
- No preservatives
- No MSG

To our knowledge, neither the sugar lobby nor the makers of Saccharin, NutraSweet, or the other substances mentioned have sued over these common labels or claimed that they mislead consumers.

The back of the packaging is an almost equal goldmine of facts and figures, including a list of the candy's ingredients and the ubiquitous (in the United States) "Nutrition Facts" chart. Yet in the United States, where most genetically engineered food is grown and sold, it is not possible to find out if GM food is one of a food product's ingredients, with the exception of a small number of highly restricted labels for growth hormone in dairy products.

In this chapter, we are going to have a look at the pivotal issue for most people in the GM food debate: their right to know what they are eating.

LABELLING IN THE US
How do you know if the food you are eating is genetically engineered? The tomatoes in your salad and the oil in your frying pan are both at risk. Even your soft drink could contain genetically engineered corn syrup. Without labels to tell you if a product is genetically engineered or has genetically engineered ingredients, you simply don't know what you're eating.

A great amount of attention has been paid to the issue of labelling because your right to know is at stake. At the heart of the labelling debate is the struggle between a corporation's right to "commercial speech" and citizens' fundamental right to know what they are buying and eating. The decision to allow the public to consume unlabeled genetically engineered food strikes some people as grossly undemocratic and slanted too far in favour of corporate interests. Should our society allow the purported commercial rights of a corporation to supersede the citizen's right to make informed decisions in the marketplace? Rather than allowing multinational corporations to experiment with our health and the health of our families, consumers should demand to have genetically engineered food labelled. The responsibility (and liability, when health/environmental problems occur) to prove safety should rest with the agribusiness giants who create genetically engineered food.

Instead, we are in the unenviable position of having an untested technology thrust upon us, and we have to take the responsibility to prove safety ourselves. The public should not have to bear this burden or the cost of safety and environmental testing, especially since we never asked for the technology in the first place and do not benefit from it in any way.

Eating organically grown foods is currently the best way to avoid GM food. This option is not foolproof, however: even organic growers and backyard gardeners can be duped into buying genetically engineered seeds because many seed packages are not well labelled either. In fact, in Monsanto's 1998?1999 New Leaf Product Guide and Seed directory, the term "genetically engineered" never appears. Farmers unfamiliar with genetic engineering may not even realise what they are buying, growing, or putting on supermarket shelves. As we mentioned in chapter 2, this New Leaf potato is considered a pesticide and, like a can of Raid, is regulated by the EPA. Without labels, you could be growing these toxic taters right in your own backyard.

Furthermore, because of horizontal gene transfer, organic fields can become contaminated by genetically engineered pollen from nearby fields.

Such contamination can occur via the wind, from pollen stuck to bees, or even when neighbour farmers share equipment. Because of this contamination, even a farmer who uses organic seed and follows organic standards perfectly can still be unwittingly selling genetically modified crops. Only expensive, sophisticated tests can reveal the contaminant DNA.

Food producers often use soybeans or corn from many different sources and growers. In the United States, genetically engineered crops have not been segregated from normal crops, and therefore many food producers cannot tell consumers or supermarkets if their product contains genetically engineered ingredients.

Currently, the FDA does not require growers, food manufacturers, or seed sellers to label their products as genetically engineered. It is a purely voluntary system for which, as you can imagine, there have been few takers. Agribusiness corporations know from public opinion polls that demand for genetically engineered food is low and that demand for labels is high. Labelling could translate into commercial failure for genetically engineered foods – precisely the reason that biotechnology companies and agribusiness giants are trying to keep labels off their genetically engineered food products.

Shoppers in some European countries and to a limited extent in Japan can find labels with information about genetically engineered content. This debunks complaints from US GM food manufacturers that labelling is impractical and too expensive: what they claim is impossible in the United States they do every day across the ocean.

VOLUNTARY LABELS

The FDA will not "require things to be on the label just because a consumer might want to know them."
James Maryanski, Biotechnology Coordinator, FDA

The FDA has reluctantly permitted voluntary GM food labelling since 1992. A few companies have opted to label their foods – mostly organic growers or those in the health food industry – but the policy is ineffective for regulating genetically engineered food products because the vast amount of food sold in the United States does not have its GM food content labelled.

A voluntary food labelling policy is a mild first step, permitting companies to provide information that their customers want. But a voluntary system creates an inequitable food-production system. One party, such as a seed seller, may label its seeds, but until there is an equitable system, information about genetically engineered seeds may never reach growers, processors, sellers, or consumers. Until all parties at every stage of food production are informed, voluntary labels remain grossly inadequate.

REVOLVING DOORS: AGRIBUSINESS REGULATES ITSELF
Monsanto, which makes large donations to both the Democratic and Republican parties and to Congressional legislators on food-safety committees, has become a virtual retirement home for members of the Clinton Administration. Trade and environmental protection administrators and other Clinton appointees have left to take up lucrative positions on Monsanto's board, while Monsanto and other biotech executives pass through the same revolving door to take up positions in the administration and its regulatory bodies.[2]
Toronto Globe and Mail, February 20, 1999

The ability of the Food and Drug Administration to regulate genetically engineered food products was tested during the approval of Monsanto's rBGH product, Posilac. During the approval process the FDA developed a labelling policy that strongly discouraged dairy farms from labelling their products as "rBGH-free." Given the controversy over the human and animal health effects of rBGH, the state of Vermont passed a law requiring labels on all dairy products that contained rBGH. The Vermont law protected consumers by informing them of what was in their milk and offering a clear alternative to milk from cows treated with rBGH. Monsanto filed a lawsuit against the state of Vermont claiming that their constitutional rights were being violated because of Vermont's labelling law. It was reported that "while working for Monsanto [Michael] Taylor had prepared a memo for the company as to whether or not it would be constitutional for states to erect labelling laws concerning rBGH dairy products. In other words, Taylor helped Monsanto figure out whether or not the corporation could sue states or companies who wanted to tell the public that their products were free of Monsanto's drug."[3] Coincidentally, it was the same Michael Taylor who was working for the FDA at the time that Monsanto's rBGH product was undergoing FDA approval. Even more curious is the fact that Michael Taylor, a former Monsanto employee, was made responsible for developing FDA labelling policy for rBGH. With Taylor's help, the FDA declared that milk from cows treated with rBGH was just like regular milk, and later Taylor returned to work for Monsanto.[4]

The connections between the US FDA and large agribusiness corporations are disturbing. If the FDA is serving corporate interests, then who is serving the public? rBGH is a classic example of regulatory corruption. Michael Taylor was not the only FDA official with connections to Monsanto: others who formerly worked for the company include Margaret Miller, in veterinary medicine, and Suzanne Sechen, who reviewed Monsanto's rBGH application.[5] Betty Martini, a consumer representative, says, "The Food and Drug Administration is so closely linked to the biotech industry now that it could be described as their Washington branch office."[6] Many other countries have banned rBGH, including Ireland, Canada, Great Britain, the Netherlands, France, Belgium, Luxembourg, Spain, Portugal, Italy, Germany, Austria, Switzerland, Norway, Sweden, Finland, Denmark, Greece, New Zealand, Australia, and Israel.[7] The US FDA actions in support of rBGH are not only out of touch with consumer demands, but with the rest of the world.

rBGH-FREE LABELS

As US consumers watched rBGH pass through the FDA with unusual ease, Ben & Jerry's, an ice cream company based in Vermont, vowed not to use rBGH and to label their products as "rBGH-free." However, the ice cream company found that certain US states, such as Illinois, had laws prohibiting dairy manufacturers from saying that their products were rBGH-free. Since nationally distributed dairy products cannot have labels tailored to each individual state, the law in Illinois prevented Ben & Jerry's and other dairy producers from labelling their products as "rBGH-free." Ben & Jerry's, along with Stonyfield Farms, Whole Foods Market, Inc., and Organic Valley dairy farms, filed a lawsuit against the state in 1997, and Illinois was forced to reverse their labelling laws, as a federal court settlement found that the law violated First Amendment rights by denying consumers important information. Perry Odak, CEO of Ben & Jerry's, said, "We regret that the state of Illinois forced us into legal resolution of this matter. However, this is a fundamentally important issue. Manufacturers should be able to tell the customers how their products are produced and consumers should have a right to information that allows them to make an informed choice."[8]

As it stands now, dairy producers can label their products as "rBGH-free" but the FDA requires that they also print a qualifier, which weakens the "rBGH-free" label and continues to be a source of controversy. For example, the Ben & Jerry's label reads:

> We oppose Recombinant Bovine Growth Hormone. The family farmers who supply our milk and cream pledge not to treat their cows with rBGH. The FDA has said no significant difference has been shown and no test can now distinguish between milk from rBGH treated and untreated cows.[9]

In addition, citizens groups such as Mothers & Others and Rural Vermont have created lists of dairy producers who have pledged not to use rBGH.[10]

INDUSTRY'S VIEW
Trade is the future of agriculture so anything that impedes freer trade is troublesome.[11]
Linda Fisher, VP for Federal Government Affairs,
Monsanto, May 15, 1997

The multibillion-dollar, multinational leaders of the agribusiness industry attempt to defend their anti-consumer, anti-labelling stance by claiming that the costs of labelling are just too high. For companies like Monsanto, Novartis, and Dupont, whose sales were $8.648 billion,[12] approximately $21 billion,[13] and $24.767 billion[14] in 1998, respectively, the claim of poverty is difficult to believe. The *entire* US organic industry reported sales of $4 billion in 1997, yet it manages to segregate crops and label foods at every stage of production. Some organic food costs more, but the costs do not arise from placing a few extra words on labels and shipping crates; rather, the price of organic food reflects high labour costs and the quality inputs necessary to produce organic food.

Sometimes industry supporters argue that consumers are just hysterical and ill-informed. On the contrary, some of biotech's biggest critics include molecular geneticists, cell biologists, lawyers, doctors, consumer groups, and farmers. These people are not hysterical, ill-informed, or anti-science, but simply critical of the crude techniques used to produce GM food, as well as the potential environmental and human health risks. Gillian K. Hadfield, a professor of law at the University of Toronto in Canada, writes, "It's wrong to view consumer resistance as just anti-science hysteria. Many people make food choices based on ethical considerations, deciding not to eat veal, or mass-produced chickens, or non-organic produce. If biotechnology raises ethical and environmental concerns for them, it is not irrational for them to act on these."[15]

SUBSTANTIAL EQUIVALENCE

> In the planting of genetically changed crops around the world, the US government has done just about everything it can to help except drive the tractor.[16]
>
> Bill Lambrecht, St. Louis Post-Dispatch

One of the ways in which the FDA and other regulatory agencies evade labelling is by applying the principle of substantial equivalence to genetically engineered food. The industry has declared that genetically engineered food is "substantially equivalent" to normal food. According to this principle, the FDA considers Monsanto's New Leaf potato to be the same as a regular potato. While the "substance" that makes up New Leaf potatoes is radically different from normal, non-toxic potatoes, regulators have apparently chosen this approach to promote the GM food agribusiness industry. In doing so, however, the agency charged with protecting citizen health and welfare has decided in advance not to conduct the prudent steps toward protecting public health and safety that would in fact fulfil the agency's mandate.

The theory of substantial equivalence allows corporations to produce novel food products and have them treated the same way that traditional, pure foods are treated. The advantage this gives agribusiness is enormous. Several respected scientists, writing in *The Ecologist*, say:

> The genetically engineered food could be compared with any and all varieties within the species. It could have the worst characteristics of all the varieties and still be considered substantially equivalent. It could even be compared with a product from a totally unrelated species or collection of species. Worse still, there are no defined tests that products have to go through to establish substantial equivalence. The tests are so undiscriminating

that unintended changes, such as toxins and allergens, could easily escape detection. For example, a genetically engineered potato, grossly altered, with deformed tubers, was nevertheless tested and passed as substantially equivalent.[17]

LABELS FROM A TO Z

We have many labels on food products in the United States. We know if our meat is kosher, if our tuna is dolphin-free, if our chicken is "free-range," and if our vegetables are organic. We know if the food is grown in the United States, or even if it is made with unionised labour. We know if it contains NutraSweet, if it is low-fat, imported, or low-cholesterol. With the wide variety of labels on food packages today, many people assume that they are getting the full story. However, many food products today contain genetically engineered ingredients, and they are unlabeled.

Philip Bereano, professor of technical communication at the University of Washington, Seattle, writes, "The failure of the US government to require that genetically engineered food be labelled presents consumers with a number of quandaries: issues of free speech and consumers' right-to-know, religious rights for those with dietary restrictions, and cultural rights for people, such as vegetarians, who choose to avoid consuming food of uncertain origins. Some genetic recombinations can lead to allergic or autoimmune reactions. The products of some genes which are used as plant pesticides have been implicated in skin diseases in farm and food market workers."[18] In 1997 the New York Times had Genetic ID of Fairfield, Iowa, test several foods to determine whether or not they contained genetically engineered ingredients. They tested four soy-based baby formulas and eight other products made with soy or corn. Most of the products contained hidden GM food. "The formulas – Carnation Alsoy, Similac Neocare, Isomil and Enfamil Prosobee – all tested positive. Eden soy milk tested negative. Morningstar Farms Breakfast links and Morningstar Farms Better n' Burgers, Betty Crocker Bac-os bacon bits, all soy based products, also tested positive. And so did three corn based chips – Fritos, Tostitos Crispy Rounds and Doritos Nacho Cheesier."[19]

LABELLING IN EUROPE
International politics continues to be dominated by the interests of large corporations. Monsanto CEO Bob Shapiro was one of the largest contributors of "soft money" to Bill Clinton's 1996 re-election campaign, and he later became a special trade advisor to Clinton.

The United States is also a major player in the Codex Alimentarius Commission, often referred to as Codex, an international group that sets food labelling standards. Codex is an agency of the United Nations World Health Organization and the UN Food and Agriculture Organization. Codex is comprised of the world's national food regulators – regulators that are supposed to have the public interest as top priority.

The Canadian and US delegations to Codex meetings have openly opposed the labelling of genetically engineered food. It is not surprising that nine of the thirteen non-governmental Canadian representatives to Codex represent large agribusiness corporations. In the United States, representation is just as skewed, since ten of the fourteen non-governmental spots are filled by large corporations or industry groups. Many corporations have grabbed a spot in both the Canadian delegation and the US delegation, including Nestle Foods, Bestfoods, Inc., and Procter & Gamble. Other corporations involved in setting Codex standards include Monsanto and Mead Johnson and Co. Codex meetings have been dominated by multinational agribusiness, while the agency slowly proceeds through an eight-step process to create international food standards.[20]

Many countries have not waited for Codex to issue their

international standards. Instead, those countries have taken the necessary steps to protect their citizens from the potential dangers of genetically engineered food. The British Medical Association (BMA) published a report, "The Impact of Genetic Modification on Agriculture, Food and Health"[21] in 1999 that called for caution and demanded food labels. "The BMA, concerned about health risks such as allergenicity, called for a moratorium on planting [genetically modified] crops until there is scientific consensus on the long-term effects of GM products. In its report, the BMA also said that if GM foodstuffs, such as soya, are sold to the public, they should be separated from non-GM foods and clearly labelled."[22] In fact, British law currently requires that genetically modified foodstuffs be labelled.

As early as 1997 European supermarkets began demanding that suppliers use non-genetically modified ingredients for their food products. For example, Sweden's Konsum supermarkets asked suppliers "to guarantee that the products they supported did not contain any genetically altered materials."[23] Tesco supermarket stores, a British chain with over seven hundred stores, has worked to find suppliers who do not use genetically engineered food products. The *Washington Post* also reported that Nestle UK Ltd. "would ensure that the vast majority of its products will not be made with genetically modified ingredients – and it would label any products that have such ingredients."[24] Other British companies soon followed suit: "Birds Eye Wall's and Van der Bergh foods, two popular brands of processed foods owned by the consumer products giant Unilever UK, declared their intention to stop using, "for the time being," genetically modified ingredients."[25] In April 1999 seven European supermarkets formed a consortium to eliminate GM food from their production system. Members include J. Sainsbury Plc (Britain), Marks & Spencer (Britain), Carrefour (France),

Delhaize (Belgium), Effelunga (Italy), Swiss Migros, and SuperQuinn (Ireland).[26] In June 1999 the European Union (EU) passed a moratorium on genetically engineered food field trials. In India, the Supreme Court also banned the testing of genetically engineered crops, and activists have literally torched crops that were genetically modified.[27]

AN INFORMED CHOICE?

While the US federal government has been slow to act on the labelling of GM food, individual states have begun to respond to the concerns of their citizens. In 1999 Maine became the first US state to consider an act establishing mandatory labelling for all genetically engineered foods. The Maine bill required distributors to label food products that had been genetically engineered, so consumers could make informed choices. The bill defined genetically engineered food as any "substance for human consumption that contains a genetic material from another species or a genetic material assembled in vitro, which genetic material is introduced into the substance by non-sexual means as the result of a current or previous application of a recombinant deoxyribonucleic acid, or rDNA, technique or other similar technique for genetic manipulation capable of combining or introducing genetic material from dissimilar organisms." Excluded from the definition of genetically engineered foods were those "developed exclusively through traditional methods of breeding, such as artificial insemination, embryo transfer, hybridisation or non-directed mutagenesis, nor does it include foods containing extracted products of a genetically engineered organism with no more than trace amounts of the organism itself or its genetic material."[28] Unfortunately, the bill died in the legislature, but more are sure to follow.

Since international negotiations appear to be corporate dominated, and national politics are clouded with agribusiness campaign financing and the "revolving door" syndrome at regulatory agencies, labelling of GM food in the United States might best be achieved in the near term at a local level. And because the United States leads the world in genetically engineered food production, the enactment of useful, complete labelling of GM food in the USA will benefit people worldwide.

There's an old saying, "what you don't know can't hurt you." Unfortunately this bit of folk wisdom predates hidden changes in our food. If genetically engineered food is safe, and people who buy food in supermarkets are not hysterical and ignorant, then the GM food makers should have no reason to oppose labels that reveal the product they claim to be proud of.

Putting food into our stomachs every day puts us at the mercy of those whom we pay to provide our sustenance. Every person has a right to safe and healthful food. Every person has a right to make choices about what they eat.

Every person has a right to know.

5
FOOD FIGHTS

> While we sit at our computers editing our copy, sending our e-mail and expressing our cyber freedoms, the TNCs [transnational corporations] are using their global networks (fed by far greater resources) to achieve concrete results.[1]
>
> Jerry Mander, author, ecologist, activist

People feel passionate about their food. Our relationship to our sustenance is tied up with our feelings of security, personal satisfaction, and cultural or ethnic identity. Most parents respond very strongly to the task of putting enough good food on the table to feed their children. People who grow, sell, or care about food are used to hearing deeply held and vigorously expressed opinions: each human being is involved every day in eating.

We are all food experts.

At the same time, the transnational corporations Jerry Mander refers to at the start of this chapter are, in their way, equally passionate. The genetic engineering companies have large – in some instances astoundingly large – amounts of their money tied up in producing genetically engineered food. Charles Benbrook, an analyst of food-production systems, has pointed out that companies like Monsanto took on great amounts of debt as they went on a buying spree, acquiring seed companies at the end of the twentieth century. He writes, "Monsanto spent $8 billion in the first half of 1998 buying seed industry assets...To cover its bets thus far, Monsanto must increase profits by $1 billion to sustain satisfactory returns to investors."[2] A good return on the investment in seeds and agriculture is a must for these corporations to thrive, and in some cases, survive.[3]

The Monsanto corporation has 100 percent ownership of three soybean seed companies, three corn companies (including the number two company and another that has over 35 percent of US corn acres), two cotton seed companies, one potato seed company, the largest US tomato producer, and a vegetable oil seed company.[4] Monsanto has a lot to lose; it's easy to see why they have been one of the most aggressive in defending and protecting their position in the marketplace.

Under these circumstances, agribusiness corporations are not inclined to take criticism or what they see as threats to their business lightly. On the other hand, citizens also express powerful opinions, especially when they feel that their access to good information about safe food has been denied.

How do we know people care about GM food? In 1997, Novartis, manufacturer of genetically engineered foods as well as Gerber baby food, released a poll in which "93 percent of respondents agreed that labelling is needed."[5] A more recent *Time* magazine poll shows that 81 percent of those asked believe that genetically engineered food should be labelled.[6]

TYING FARMERS' HANDS
One group of people who have directly experienced the power of biotech corporations is American farmers. Longtime traditions such as saving seeds, sharing seeds with neighbours, and even selling surplus seeds have been

sacrificed in order to meet the legal requirements that are a mandated part of growing genetically engineered crops.

Legal requirements? What kind of legal requirements would farmers need to farm? Farmers don't really need legalities to grow our food, of course, but giant agribusiness companies require some farmers to sign legal documents compelling them to grow only that company's seed, use only that company's chemicals, and pay "technology fees" for the genetically engineered seeds in addition to the cost of the seeds themselves. Some legal agreements allow corporate representatives independent access to the farmer's land, so they can take samples of crops back to the lab to see if the farmer is cheating on the agreement. And the documents impose penalties for any failure to meet the contract terms. Percy Schmeiser, a Saskatchewan farmer, was sued by Monsanto in 1999 for, as Mr. Schmeiser put it, "doing what I've always done." His mistake? Saving seeds. Hundreds of American and Canadian farmers have been sued by Monsanto for saving seeds.[7] Monsanto will go after individual farmers because the seeds, they say, are theirs, protected by US patents. The biotech firms claim rights to not just the first seeds sold, but all future generations of the plants too. If Monsanto, Novartis, Zeneca, and other big biotech companies had it their way, farmers would buy seeds from them over and over again.

There is another reason for the zeal to force farmers to buy a suite of products. While most biotech companies are staunch defenders of the patent system and the limited monopoly rights they can gain from that system, even patents run out. Generally, when a patent runs out the item under patent can be made and sold by other companies. The assumption in our patent laws is that the monopolistic protection afforded by the patent allows the originator of the invention to recoup the investment it took to think up and develop that novel product. Then the free market is supposed to take over.

Monsanto's herbicide Roundup has been a real cash cow for the company, with sales of more than $2 billion yearly.[8] Glyphosate-based herbicides like Roundup are very common, ranked third for use on industrial and commercial land. The EPA estimates that farmers use between 38 and 48 million pounds of glyphosate each year.[9]

The problem is that Monsanto's patent on this product is set to expire in the year 2000. Rather than lose the income from the monopoly, what if Monsanto could simply *require* farmers to use its product? It would be as if Ford owned Exxon and made everyone who bought a Ford sign a contract obliging them to stop only at Exxon gasoline stations. That is just what Monsanto has done with its Roundup Ready line of products, including Roundup Ready corn, cotton, soybeans, and canola-crops that are grown on millions of acres around the world. By requiring farmers to use Roundup on these crops instead of other glyphosate-based herbicides that will be available, Monsanto has effectively extended the monopoly aspects of its patent beyond the expiration of the patent itself.[10]

While farmers are technically free to plant what they choose, the realities of seed marketing limit the choices that farmers can make in some locales. So they sign, buy the required chemicals, plant and spray, and hope that the higher costs of this arrangement will pay off in higher yields. So far, the results of this kind of bio-contract agriculture have been at best mixed, with increasing signs that many if not most farmers will not see those higher yields translated into higher returns, as they find fewer and fewer outlets for their engineered produce. Ironically, some buyers are now offering a *premium* for conventionally grown produce, making the economics of genetic engineering even riskier.[11]

DON'T INSULT YOUR SALAD
In 1997 Oprah Winfrey, on her national TV show, held a

Food Fights

special programme about the cattle business. After talking with guests about the beef industry, she made the now infamous statement, "I'll never eat a hamburger again." For that comment, the beef industry sued Oprah in a highly publicised court case, heard in Amarillo, the heartland of the Texas cattle industry. By voicing her opinion, the beef industry said, she had damaged their business and broken the law. The law she had broken? One that prevents a person from publicly criticising food products.

Veggie Libel Laws, or Food Disparagement Laws as they are more officially known, are on the books in thirteen US states. These laws make any statement that might diminish the income of food producers by disparaging a food product in public subject to court action. Our constitutional protection of free speech notwithstanding,

US STATES WITH FOOD DISPARAGEMENT LAWS IN 1997

Alabama	Mississippi
Arizona	North Dakota
Colorado	Ohio
Florida	Oklahoma
Georgia	South Dakota
Idaho	Texas
Louisiana	

legislatures in state after state have passed laws limiting the right of citizens to speak out on issues that affect them.

Few constitutional scholars can be found who believe that such laws would be upheld if taken to the US Supreme Court. But constitutional or not, the veggie libel laws are very effective for two reasons.

First, a person expressing an opinion about her salad in

public might think twice about the possible cost of defending herself against a veggie libel suit, even if she won. Oprah Winfrey had to stay in Amarillo throughout the lengthy trial and spent over one million dollars in legal defence fees.[12] Ms. Winfrey is one of the wealthiest women in the United States. What about ordinary people who might want to speak out about food? The result of the first factor, the high cost of defence, leads us to the second factor, what people in the civil liberties field call the "chilling effect." Faced with the possibility of a lawsuit won in the courtroom and lost in the chequebook, it would be understandable if a person considering speaking out about food issues decided not to return a reporter's call, not to appear on a public platform, or not to write a letter to the editor. For those whose goal is to shut off public debate about food issues, the cost of lobbying and exemplary lawsuits can be an excellent investment if it helps to turn the spigot of public discourse about their issues to the closed position.

In 1998 scientist Marc Lappé and activist Britt Bailey were set to publish their book *Against the Grain*, which provides a good deal of information, some of it significantly critical, about some food biotech firms. Just before going to press the publisher, Vital Health Publishing of Bloomingdale, Illinois, received a letter from the Monsanto corporation threatening legal action.[13] The publisher, fearing ruin in the courtroom, dropped the book. Some months later the pair found a heroic if smaller publisher, Common Courage Press, which brought out the book. Monsanto took no action. Perhaps their purpose had already been served. The book was delayed and kept away from a larger publisher.

Similarly, in 1998 the prestigious British magazine *The Ecologist* learned that its special issue on the Monsanto corporation had been destroyed by its long-time printer, Penwells of Saltash, Cornwall, just before distribution but after printing. The printer indicated that it had heard from Monsanto and was afraid of leaving itself open to libel action. Eventually another printer was found and the issue's contents were made available publicly, over the internet and on paper. The "Monsanto Files" was *The Ecologist*'s most popular issue, printed in six different languages, with over 300,000 issues sold worldwide and no indication of a slowing demand.[14]

This threat imposed by large biotech companies upon the media is not an uncommon occurrence. For example, in 1998 a team of well-respected reporters, Steve Wilson and his wife, Jane Akre, was aggressively investigating Monsanto's bovine growth hormone. Employed by Fox TV, the two reporters were working on a series examining the politics, science, and social consequences of rBGH. Immediately prior to the airing of the program, Monsanto threatened to sue Fox for reckless, biased presentation of unscientific speculation. According to Wilson, "Fox attempted to cover up the truth by firing us and then having a newly hired, less-experienced reporter redo the series leaving out crucial facts and reporting some of the same lies and distortions we refused to broadcast. It wasn't just what he left out, it was what he left in that makes his piece so egregious."[15] This censorship by a large, powerful corporation puts our free press at risk.

While writing this book, one of the authors was telephoned by an Emmy-Award winning television producer who was preparing a story about genetically engineered food. The producer subsequently called, at our suggestion, a major food biotech firm to hear their side of the issue. In that conversation the producer reported being threatened three times. Subsequently, the producer dropped that company as the central example in the planned TV story.

FULL COURT PRESS
It would be a mistake to think that the legal system is

available only to wealthy corporations bent on squashing dissent, although those with in-house legal teams and big budgets certainly enjoy some advantages. In recent years a number of non-profit organisations have followed the pioneering work of Wharton-trained economist Jeremy Rifkin, whose organisation, the Foundation on Economic Trends, has filed not only lawsuits but petitions and other legal actions designed to hold the government to its role of protecting the health and welfare of the citizenry. In 1997 the environmental activist group Greenpeace, joined by thirty other organic farming organisations, filed a petition with the EPA to cancel its approval of crops engineered to contain the natural pesticide *Bacillus thuringiensis* (see more on Bt in chapter 6.

On December 15, 1998, the Center for Food Safety, along with several other citizens groups, began legal action against the FDA concerning their approval of rBGH. Other lawsuits and petitions have been filed in recent years on the labelling of genetically engineered food and Bt crops, and more recently on genetic pollution. While few of these legal actions have resulted in judgments in favour of the citizens groups, at least so far, such legal strategies serve other purposes as well.

A crucial goal in the filing of suits and petitions is to focus public attention. Few of these actions are sent to courts or agencies without a press release. Perhaps because the media seem attracted to the appearance of conflict that legal actions suggest, such filings are often well covered in the press. Thus the public, the government agencies, the courts, and the legislative branch all become better informed.

Because legal campaigns are often costly to mount, even with low-cost or pro bono attorneys, non-profit citizens groups band together to share resources. This helps to build movements for change by bringing groups that might otherwise be competing into cooperation and coordination.

Food Fights

The occasion of the petition or suit might also be a time when diverse activists sit down and hammer out a shared position on a given issue, yielding a more coherent social movement.

Finally, such coordinated actions help to strengthen democratic institutions by holding government accountable and asking for redress of grievances. Big biotech companies have major financial resources to deploy to purchase or influence media time, to contribute to political campaigns (something non-profit tax-exempt citizen's groups are forbidden to do in the United States), and to finance expensive studies by hand-picked researchers. When citizens band together and speak to their government, sometimes the government listens. Other times, it takes a judge to help get the government's attention.

In December 1997 the USDA proposed new organic standards that would have allowed genetically engineered food to be labelled "organic." Scores of citizens groups sounded the alarm to their members and constituencies, and comments opposing these proposed rules flooded into the Department of Agriculture by the hundreds of thousands in what Agriculture Secretary Dan Glickman himself called "an absolute firestorm."[16] The proposed rules were withdrawn.

In June 1999, at a national summit on genetically engineered foods in Washington, D.C., a petition was presented to Congress and the FDA with over 500,000 signatures. The petition called for the government to make the labelling of genetically engineered foods mandatory.[17] The debate over GM food is emotional and sometimes loud. In an era of powerful, confident, media-savvy giant corporations, citizens concerned about genetically modified food can benefit from working together, using the legal and media systems well, holding their governments accountable, and maintaining their courage.

6
FIELDS OF GREEN: FARMING AND BIOTECH

The rattling of a shopping trolley is a lot more familiar to many of us than the roaring of an old Ford 8N tractor. Although most of us do not grow the food we eat, our connection with farming is powerfully simple: 60 percent of the daily human caloric intake of plants comes from just three staples – wheat, rice, and maize.[1]

Farming is not the romantic back-to-the-land pursuit described by some modern nature writers. It is hard work that is often not rewarded with respect, results, or a decent income. Yet without our farmers, we would quickly die, since only a tiny number of people in our society know how to grow their own food, and even if they do know, few have the land, tools, water, and proper seeds to keep starvation away. What this adds up to is a complete dependence on a small group of people whom most of us don't know and never see. Even without ever encountering a farmer, we have a powerful interest in protecting and preserving farming, since it sustains us all.

Reminding ourselves of our close connection to agriculture is important if we are to understand the true significance of the genetic revolution in agriculture. In this chapter we're going to see what that connection is and what it means for humanity, and why genetic engineering poses the deep threat that it does to our ancient system of supplying ourselves and our communities with food to eat.

TWENTY-FIRST CENTURY FARMING
Even if glossy packaging and brightly lit supermarkets fool us into forgetting our daily connection with plants, soil, and water, we are nonetheless intimately part of the growing of food – each one of us, every day, no exceptions.

In hundreds of centuries of agriculture we have been transforming the physical features of our globe, chiselling away at mountainsides, changing the course of rivers, building up and tearing down the living soil all around the world. As we chop down rain forests and pipe water deep into deserts, we are fast approaching the time when all land that can be used to grow food has been found, transfigured, and put into production. As we run out of places to grow food, we are pouring roads and building foundations over former farmland, strip-mining the nutrients from our soil, polluting our groundwater and soil, and filling huge tracts of land with salts so that nothing will grow there anymore.

Our impact is biological, not just physical. It is estimated that in the year 2000, over two billion tons of rice, wheat, maize, and barley, among other crops, will be consumed across the world, a 25 percent increase from 1995.[2] We and our animals eat 3 percent of all the plant matter on earth each year.[3] If you add in everything we cut down, pave over, or otherwise eliminate, human beings consume an incredible two-fifths of all the plant matter on the earth yearly.[4] We are not a low-impact species.

Meanwhile, the number of people classified as farmers in the United States has fallen to such a small number that the category "farmer" was slated to disappear from the country's census of its citizens in the year 2000. The popular image of the self-sufficient farmer's traditional mix of livestock, vegetable gardens, and crop fields is no longer a reality. It has given way to a highly specialised, technology-based industry. While we tend to frown upon the ecological destruction and dehumanisation wrought by such vast modernisation, industrial agriculture lies at the heart of contemporary society, whether we recognise it or not. Favouring low costs, individuals and industries alike promote increased productivity and technological change.

HOW AGRICULTURE WORKS
WHY IT IS WORTH SAVING

The great miracle of agriculture is that, at its core, human beings shape the world with their minds. Before agriculture, people collected the wild growing food that they needed. Then, it is thought, they began to plant seeds of the plants that they favoured. It didn't take long for people to notice that if they saved the seeds from the biggest or sweetest or strongest plants, the next generation of plants would tend to have that characteristic. And when the best food was saved from that daughter crop, the next generation was even better.

Thus humanity, by selecting what we liked to eat, gradually shaped food to our liking – making real our mental picture of what constitutes good food. What an incredible, subtle, and close partnership this has been. The plants keep us alive, we keep the plants alive. We nourish each other; we grow up and grow old together. We adapt together, we travel to the corners of the earth together. And in times of famine or catastrophe – we die together.

Our food plants are not just the result of individual mental pictures. They are the creations of human communities, and they are mostly the creations of women.[5] When a farmer looks over a field and chooses which plants to save seeds from, she is putting into action the values and culture of her entire community. Food tastes are a powerful aspect of community coherence and identity, so much so that when we eat tomato sauce we think Italian, and when we eat water chestnuts, Chinese. One of the great threats of genetically engineered food is the disruption and destruction of these essential connections between culture and food, people and plants.

GM food brings with it several features that impair the human relationship to food, especially ownership of life, corporate control, and bio-piracy. In coming to an understanding of the menace of genetically engineered food, we will look at how each of these GM food features arises and whose interests it serves.

OWNING LIFE
Some years ago one of the authors was visiting a friend in a remote village of southeastern Nigeria. Stepping out of the car, he was greeted by a horde of laughing, naked children who instantly grabbed his camera, notebook, pen, and wallet. He watched with slack-jawed disbelief as his American Express card bounced off into the jungle in the grubby fist of a small boy, his passport flying off in another direction amidst a throng of giggling little girls. The British nurse who lived there put a comforting hand on his shoulder. "Don't worry," she said. "People who don't own anything can't steal." Later on in the day, children filtered into the nurse's little house and dropped the contents of the visitor's pockets into his lap, one by one, their curiosity satisfied. The visitor asked about the downcast expression he noted on many of the children's faces. The nurse explained that the children knew that the man thought he owned all of that stuff; this aberration embarrassed them.

In all Western societies some regime for ownership exists, both in custom and in law. A great deal of attention is paid to setting up clear mechanisms for establishing and enforcing who owns what. Titles, deeds, receipts, and a huge body of law confirm possession. Basic to the idea of ownership in Western countries is the patent system. In this aspect of ownership, the government grants exclusive rights to an invention in exchange for registration and a promise to later on make that invention available to one and all.

The patent system is so important in the United States that it is established in Article I of the US Constitution. The entire US patent and copyright system is established in these twenty-seven words: "To promote the progress of science and useful arts by securing for limited times to authors and inventors the exclusive right to their respective writings and discoveries."

Over the years agriculture had little involvement with the patent system because as objects of nature, plants and seeds were thought to be outside the scope of patent law. Further, there wasn't even much of a seed industry in the United States until relatively recent times: people saved their own seeds or traded and bartered with neighbours. In an agrarian culture this was natural and also good agricultural practice, because obtaining seeds from neighbours helped to insure that the resulting plants would be best adapted to local growing conditions and would produce crops that fitted local tastes.

While a few seed companies existed in the 1800s, according to Cary Fowler, a former senior officer at the United Nations Food and Agriculture Organization, the main source of seeds in the United States through much of the nineteenth century was the federal government, which distributed millions of seed packets nationwide. The goal was the creation of a stable and diverse agricultural system.

As seed companies began to grow, laws were passed to protect their "innovations." The US Congress passed a law moving toward ownership of agricultural plants and seeds in 1930, permitting the classifying of some plants as a form of intellectual property like books and music. In the United States and throughout the industrialised world, these developing concepts of plant ownership, known generally as "plant breeders' rights," granted exclusive control over certain plant varieties for up to twenty-five years.

The system of plant breeders' rights led to the next big

step, granting patents over life. Philip Bereano says, "For 200 years the idea that general patents could cover life forms was viewed as ridiculous."[6] Following a little-known Supreme Court decision in 1980, the way was cleared for the owning of life. Few people to this day know that ownership rights have been granted not only over plants but also over seeds, microorganisms, genes (of plants, animals, and people), various proteins, and even entire species.

At the same time that laws were being passed in industrialised countries giving corporations proprietary rights over living things, other laws were created to link up all these patent regimes in a scheme called "patent harmonisation." Operating principally though sweeping international trade agreements, this system of harmonisation not only yokes ownership regimes into a worldwide net of enforceable proprietary rights, but also compels smaller and non-industrial countries to establish Western-type life-ownership systems – or risk losing out on participation in the world trading community.

The key to much of the life ownership epidemic is biotechnology, because it is frequently the insertion of a gene or some other genetic engineering feat that forms the basis for the patent claim. Poor countries are unlikely to develop their own biotechnology because it requires massive capital investment. Once wealthy multinational corporations acquire monopolies on seed lines, those poor countries may become dependent upon seeds from foreign sources as they stop using traditional varieties that then go extinct. Thus, biotechnology promotes the dangerous consolidation of our global food supply as well as the exploitation and forced dependency of less-developed countries and their farmers.

The genetic engineering of food depends for its existence on the patenting of life. As we saw in the last chapter, companies that invest huge sums in the engineering of plants will only do so if they can lock up the results of their work by obtaining exclusive rights to the proceeds of those seeds.

Whatever advantages GM food brings, it is not to people who buy and eat it, but rather to the companies that build this food and, if you believe the companies, to the farmers who grow it. In addition to the problems for farmers we have looked into earlier, there are broader implications to the ownership of food. Many people believe that the ownership of life is in itself harmful. Indian physicist and biotech activist Vandana Shiva says:

> In the era of genetic engineering and patents, life itself is being colonised. Ecological action in the biotechnology era involves keeping the self-organisation of living systems free – free of technological manipulations that destroy the self-healing and self-organisational capacity of organisms, and free of legal manipulations that destroy the capacities of communities to search for their own solutions to human problems from the richness of the biodiversity that we have been endowed with.[7]

BIOPIRACY: STEALING OUR HERITAGE

As if assuming our heritage weren't enough, some GM food companies are stealing it. Biotech companies have from their early days been associated with the phenomenon known to many in non-industrialised society as bio-piracy. What is this new piracy and why does it affect us all? Food – producing plants have never been the product of one person's work.

The interaction between people and their food plants is subtle and long-range, sometimes taking hundreds of years or more to go from a tiny ear of corn the size of a finger to the foot-long sweet corn we buy in supermarkets, or from the grape-sized ancient tomatoes to the juicy behemoths that grace our summer gardens. A single variety of potato or wheat could be the result of a community's patient, collective effort over scores of generations. It is the cooperative, communal nature of plant breeding that has made it part of the common heritage of humankind.

In more recent times, plant breeding has been taken on by big universities and corporations. These production-oriented entities have searched the globe for the raw material of their plant-breeding experiments, often using what they call "plant collection." In these "expeditions," botanists, agronomists, or anthropologists travel to distant countries, usually the non-industrialised lands where the food plants originated. The collectors bring back samples to base their new varieties on. Although the plants that are "discovered" on these expeditions might be the result of thousands of years of patient community effort, the collectors just take and use. Why would they need to ask, they say, when after all the seeds or cuttings they bring home are the common heritage of humankind?

The problem arises when the universities, governments, or corporations that get the plants from other nations place strains of those plants, or processes using those plants, under patent. With the stroke of a pen (or, more likely, the press of a computer key), the common heritage of some distant people or tribe becomes the exclusive property of those who took it. Even the originators of the plant could risk infringement if they clashed with the patent holders in how they used their own foodstuffs – plants that had been passed down to them by their ancestors, plants that are intertwined with their cuisine, culture, and religion.

This is bio-piracy, the expropriation of what should never be owned, and it is possible only through the patent system that establishes "intellectual property rights" over what was formerly no one's – or everyone's – property. As Vandana Shiva points out, "Through intellectual property rights, an attempt is made to take away what belongs to nature, to farmers, and to women, and to term this invasion improvement and progress."[8]

The Seed Trade Act in Europe actually made it illegal to grow and sell "non-certified" seeds from indigenous varieties, favouring the commercial, biotech varieties. This is a good example of how far genetically engineered food has already gone in undermining small farmers and organic agriculture and furthering the loss of biological diversity.[9]

MONOPOLIES AND MONOCULTURES

The grand strategy that nature uses to ensure survival of living things is based on the web of diverse living organisms. It's a rough world out there for most plants, full of gnawing insects, creeping fungi, howling winds, and sudden killer freezes. The living things that survive over time are likely to have the most diversity in their genes, so that they can adapt most readily to the insults that chance and a competitive environment throw their way.

Plants that do not have diversity built in might fail. In the most notorious case, a million people are thought to have died in the Irish potato famine of 1845 and subsequent years. The potatoes were genetically identical, so the fungus that caused the blight infected them all. Had there been more diversity in the staple crop of the Irish people, perhaps one of the potato varieties known to have resistance to that pest would have saved many lives.

Because environments change, so can plants, especially those in the most changeable environments. Plants adapt subtly and quickly, so the same batch of seeds grown in two locations even a mile or two apart may evolve into different strains that are better suited to the conditions of soil, microclimate, and water in each area. Given just a little care, plants will do a wonderful job of optimising themselves to feed us the best and most reliable food we could ever want.

Contrast this with patented, engineered GM food. A company invests a substantial amount of money to engineer its "product." It has every possible incentive to market this crop aggressively, to recoup its investment and show a good profit. Therefore, the company will work to produce genetic uniformity – after all, the patent is based on genetically identical plants – and to keep that uniformity going as long as possible. Because of the owned nature of these GM food plants, all the marketing muscle of the patent-holder multinational corporation will go to prevent biodiversity. Yet it is biological diversity that time and ecological science have shown promotes a safe and strong environment.

By modifying plants under carefully controlled laboratory conditions and then setting up a strict regime of chemical-based cultivation in the field, companies can ensure the genetic uniformity of pure-bred, high-yield varieties. The plants that result from this process are highly invariable, not only in their appearance and yield, but also in their genes. If the genes are virtually the same, the diversity that we need to maintain ecological and planetary health is diminished. This narrowing of the genetic base has severe consequences: not only the irreversible loss of genetic make-up, but also the inability of the plants to evolve or adapt to changes in their environments.

In 1970 genetic uniformity in the US corn crop was responsible for the loss of almost $1 billion worth of corn, while yields were down by half. According to an independent report for the Food and Agriculture Organization of the United Nations prepared by the Rural Advancement Foundation International, 80 percent of the US commercial corn crop that year contained a gene that made it vulnerable to a plant disease called southern leaf blight.[10] Monoculture is dangerous, expensive, and unnecessary.

Monoculture is unnecessary because although extinction of food crops has been occurring at an alarming rate, we still have ample stocks of naturally pollinating, unpatented food crops available. One important source of heritage seeds is the Seed Savers Exchange, an organisation of backyard seed savers whose membership takes care of more than three times the variety of plants than are offered in all the commercial seed catalogues in North America combined. In its latest compilation of statistics about non-patented seed, the Seed Savers Exchange tells us that there are more than three hundred varieties of corn available in seed catalogues in the United States and Canada, many of which could harbour resistance to southern leaf blight and any number of other pests.[11] The goal is not to find one good variety of any crop, whether patented or not. Rather, what makes good agricultural and ecological sense is to plant as many varieties as we can to maximise biodiversity and minimise monoculture.

THE SELFISH ENVIRONMENT
In chapter 2 we saw how some crops are engineered to contain the pesticide Bt. This engineering approach to farming creates several environmental difficulties. One significant issue is called gene flow.

Gene flow occurs when genetically engineered genes pass to neighbouring crops. The result can be a spread of genetically engineered characteristics to other crops and weeds, posing a threat even to our ability to rely on the label "organic." Here is how it works.

When plants are genetically engineered to accept inserted, alien genes, those new genes are kept in a "turned-on" status. The inserted, engineered genes express their characteristics some or all of the time, instead of just once in a while. Further, the vectors, the biological airline that transports the inserted genes, are specially designed to move easily into new organisms – that, after all, is their purpose. So we end up with hyperactive genes carried by eager transport mechanisms. The result is a tendency of these genes to spread.

Because the genes are designed to be "on" all the time,

and because they have enthusiastic vectors for transport mechanisms, they can easily end up in neighbouring crops, even organic ones. For some time the biotech industry has insisted that pollen cannot travel very far and so genetically modified characteristics would be confined to the fields of farmers who chose to use the technology. This assertion goes contrary to the experience of worried organic growers and farmers of non-genetically engineered crops, who by 1999 were starting to fail tests for GMO-free labels in Europe.

In June 1999 the British government's Ministry of Agriculture, Fisheries, and Food, which up to that time had been an enthusiastic booster of genetically engineered food, released a report on research carried out by the John Innes Centre in Norwich, said to be Europe's leading research facility on genetically modified crops. The report showed that wind-borne pollen and bees could carry those eager genetically modified genes for miles, making it impossible to guarantee that foods sold as GMO-free were in fact GMO-free.[12]

Genetically engineered seeds can also travel, and when one looks at seeds it is easy to see why. Seeds are the vehicle for the dispersal of plant life. They employ various ingenious strategies for moving around, from developing little winglets to manifesting delicious coatings that encourage animals to carry them around or bury them for later use – where, often, the seeds just germinate and grow instead. Thus a few genetically engineered seeds dropped during planting can be picked up by the wind or by animals and carried for some considerable distance. At least one instance has been reported of genetic pollution occurring from the use of a harvesting machine that had previously been used on GM food fields.

Further, the genetic modifications can persist for a long time. When the natural pesticide Bt is used as a spray, it does its work and then degrades. When the pesticide is "turned on" all the time and is engineered into crops, the plants continue to carry the pesticidal properties long after the need has passed. University of California professor and plant expert Miguel Altieri has pointed out that once Bt crops drop their leaves or are ploughed under, the toxin remains active underground, affecting the crucial tiny life in the soil in ways that are not yet understood, but which could be highly dangerous to soil health.

FARMING AROUND THE WORLD

People in industrialised countries can become so accustomed to seeing films of huge combines moving down vast fields of identical crops that they forget about the 1.4 billion small farmers around the world who feed their own families and others in their communities. In the world's two most populous countries, China and India, small farming is the norm. What these farms may lack in size they more than make up in diversity. While life on such a farm can include a lot of hard work, many millions of people have been following this way of life with good success for millennia.

Genetically engineered crops threaten the livelihoods and even the lives of millions of these farmers. Small farmers generally have little cash. They grow what they need and they need what they grow. Therefore, the higher cost of genetically engineered seeds, along with add-ons like "technology fees," might be a slight financial factor to a large-scale American soybean farmer but a major expense to a poor peasant in India.

An even greater expense is the cost of all of the required additional elements in genetically engineered farming. Some engineered crops require a great deal more water, not always a readily or cheaply available commodity in some countries. Worse still, most GM food plants require expensive chemical amendments including fertilisers and pesticides. In one study the kind of monoculture represented by genetically engineered crops required as much as sixty times as much "inputs" of cash and labour to produce the same amount of food as traditional plant varieties.[13]

It is this huge capital investment mandated by GM food, together with the promise of a high yield, that causes small farmers to give up their diverse ways of agriculture so they can plant every square inch of their land with the biotech crop. Once they sell the crop they can use the proceeds to pay off the loans they had to take to buy all of the amendments, and use the balance to buy food. If the price for the monoculture crop drops, the farmer may not have enough money to buy food for her family. If the price drops low enough, as has happened in places like Mexico and India, the farmer stands to lose her land and watch her children starve.

Genetically modified crops bring other problems. Many of these crops require the addition of various chemicals, usually sold by the purveyors of the seed. This toxic chemical use is a threat to the health of the farmers, their families, their neighbours, and the environment, especially since there is no guarantee that the label directions for supposedly safe use will be in a language that the farmer understands – even if the farmer is literate.

The communication difficulty applies in other areas as well. Some biotech firms selling Bt crops have been telling farmers to be sure to leave sections of their land area unplanted with GM food crops to permit "refuges" for insects. These refuges purportedly permit a good percentage of the bugs to survive and reproduce without developing pesticide resistance, thus avoiding turning the species into "superbugs."

Even if the farmer can read the instructions for creating refuges, what poor or marginal farmer can afford to devote any productive land to crops that produce less income when he might already be in serious debt from paying technology fees for the biotech ones? There is evidence

Fields of Green: Farming and Biotech

that even large-scale US farmers consider the refuge requirement "a joke."[14] Given these realities, the creation of bug refuges is highly unlikely. Worse still, a University of Arizona study published in the journal *Nature* in August 1999 found that the mating cycle of Bt-resistant superbugs is different from that of non-resistant bugs, meaning that superbug populations could explode whether or not there are refuges for non-resistant bugs.[15]

Furthermore, scientists discovered in the early part of the twentieth century that our food-producing plants originated in a small number of "centres of diversity" around the middle of the globe: wheat is from Turkey, rice from China, potatoes from the Andes. These centres of diversity still exist and are the prime sources of all present and future genetic food-crop material. If as a food crop evolves it develops a weakness to a pest, plant scientists can go back to the centre of diversity for a primitive version of that crop and crossbreed the old variety with the new one, giving a kind of genetic transfusion to the crop.

Genetically modified crops threaten this basic resource. The traditional varieties are vulnerable to gene contamination from genetically engineered crops nearby, so that when agronomists or farmers seek out the old varieties to rejuvenate their crops, they might not be able to locate uncontaminated seed. The possible erosion of humanity's crop genetic heritage threatens not only farmers in non-industrial countries, but every human being.

BIO SERFS

The history of agriculture around the world is filled with the story of serfdom. From Tibet to Tennessee there are similar tales of farmers in prior centuries who worked on land owned by rich lords. These owners and overseers dominated the farmers' lives and forced them to turn over unreasonable amounts of their crops and profits for the privilege of using someone else's land. We can also see in the history of the last one hundred and fifty years the monumental and frequently successful struggles of farmers around the world to escape from the chains of serfdom in order to control their own destinies. What is called "land reform" in many places is really farmer reform, returning to farmers their autonomy, financial independence, and dignity.

Because of the biotechnology revolution in agriculture, serfdom has again reared its ugly head. In the industrialised countries, farmers no longer go into their own barn's cellar to bring out the carefully saved seed from last year's crop; they may no longer simply walk into a feed store and pay cash for a sack of seed. In many ways they no longer own their own crops, but instead simply bring the corporation's product to market and must return each year for more seeds and more rules about what they can do with their own land. Thus farmers who may have prized their independence lose control over their farming methods and even their own privacy and freedom from trespass by becoming drawn into GM food agriculture.

In non-industrialised countries the growing sense of serfdom is even more pronounced. As we discussed earlier, small farmers in countries like India who find themselves persuaded to grow genetically engineered crops may need to sign over their small farmsteads to creditors to raise the funds to get into GM food farming. The farmer's ability to act independently, especially when world commodity prices fall, shrinks rapidly.

In this age of globalisation, when market prices are determined far, far away, rapid and devastating drops in prices are probably inevitable.

Indeed, from the point of view of those who invest in globalised agriculture, such drops in prices are the main point of world-scale agriculture. If all soybeans sold for the same price around the world, there would be no need for the elaborate network of trade agreements and treaties.

People would just shop locally.

Given the good communications network enabling information about prices to be transmitted anywhere instantly, a buyer for a big grain company sitting in Minneapolis, Zurich, or São Paulo can shop for the lowest prices by computer. The buyers look for – and find – places where local conditions have caused the price to drop. There is no incentive, therefore, in the global trading system to build in price protections or stabilisation mechanisms. This deliberate instability in the world commodity pricing system makes traders rich – and farmers poor.

Because genetically engineered seeds require so much capital, and because the world trading system exploits price drops, farmers by the millions stand to lose their independence and become once again in thrall. This time, the serfs are not beholden to local lords of the manor, but rather to distant, impersonal biotech companies that don't even know the names of the farmers and families they dominate.

The hard work of farmers around the globe keeps all of us alive. Farmers deserve a fair return for their work, safe and healthy working conditions, and control of their own destinies. All of this is threatened by burgeoning biotech agriculture, and all of us stand to benefit from working to protect and defend sustainable, diverse, and healthy farming throughout the world.

7
CROSSING SWORDS WITH AN ANGEL

> We give a greeting and thanksgiving to the many supporters of our own lives – the corn, beans, squash, the winds, the sun. When people cease to respect and express gratitude for these many things, then all life will be destroyed, and human life on this planet will come to an end.[1]
>
> From the Address to the Western World
> by the Six Nations Confederacy

Pushing sautéing onions around in a favourite cast-iron frying pan, or sitting in the local pizza shack with the family, we are participating in one of the most universal of human activities. In consuming food, each one of us is a part of the continuance of agriculture, a ten-thousand-year-old invention of humankind that has altered just about every feature of human culture from how we organise our varying societies to the images in our art.

The agriculture that produces our food is crucial for all human life; the nutritional value and safety of our food are the basis for our health. In addition to these vital qualities – or maybe because of them – food holds a special place in the communal, personal, and spiritual lives of many people.

Sitting on an altar in a thatched hut in a Southeast Asian village is a handful of rice. In West Africa, yams are offered, while the ancient Greeks poured a libation to their gods in almost every tale. In front of the chancel in a modern American church is a beautiful spray of flowers – or at Thanksgiving, a basket of corn and gourds. Humans through history and around the globe have brought their gods into the subtle relationship they hold with their life-sustaining plants.

From Japan to Ireland, Paraguay to Turkey, many people begin their meals with a recognition of their gratitude for what they will eat. Individual practices might vary from family to family and from culture to culture, but the constant topic, as one surveys graces from many places

---

BLESSING

Love, we gather in your name
To celebrate all the good gifts of life:
The sun, the soil, the rain,
The seeds, the plants, the grain,
Every butterfly and bee,
And all the hands whose work and care
Have set this meal before us.
Source of Life, you have entrusted us
With the whole of creation.
May we be wise guardians of this holy place,
The earth, our home.
May we by word and deed make real
Your thirst for justice
And your infinite compassion.
Blessed be the wonders
And the beauties of this world,
And blessed be our struggle and our longing,
Love, To live our lives in peace.
Amen.

Rev. Mary J. Harrington

and times, is the connecting of our food with something that is transcendent or spiritual. An appreciation of the life that sustains us – even in its own death – is one of the profound human spiritual truths in most of our traditions.

Religious and spiritual practices concerning food are not confined to graces. Many religions have special precepts and canons governing food.

Hindus and some Adventists do not eat meat; observant Jews and Moslems do not eat certain foods and will only consume other food that has been prepared according to specific rituals. And almost all religions associate food with many of their holidays, from sweet rice cakes during the New Year in Asia to plum pudding at Christmas in England. Some religions incorporate food right into their worship, such as the Passover Seder of Judaism or the profound mystery of the Eucharist celebrated daily by millions of Catholics in Holy Communion.

Even those who are more secular in their personal practices associate the providing of food with what is most deeply human. Thus charities around the world provide food for those less fortunate, and the United Nations maintains several agencies focused on providing food for the people of the world. And when there is a disaster, bags of canned goods quickly pile up in schools, church basements, and offices of the Red Cross. Food is a universal human constant, our first and last concern for the continuation of life.

Into this picture of long history, intertwined involvement, and spiritual connection comes genetic engineering. Our deepest and in some respects most essential connection to the food that sustains all of humanity is threatened by the conversion of food from an intricate plant/culture system to a cold-blooded profit centre – the transformation of what we love and are nourished by into a product line engineered to maximise revenue at all cost. It is no exaggeration to say that bio-engineered food is a challenge to our spiritual, moral, and religious lives.

The Jewish, Moslem, and Christian traditions share one book, what is usually called the Old Testament or The Law. Comprising 53 percent of the world's population,[2] adherents of these three religious traditions hold much in common in their roots. Even for those who are not strict adherents of these faiths, a great deal of Western society has been shaped by thousands of years of Jewish, Christian, and Moslem history. At the very beginning of the Old Testament, we find the Western tradition origin story in the book of Genesis. Most people are familiar with the central role played by the Tree of Knowledge and the apple in that story. Less well known is another tree, the Tree of Life.

In the Genesis origin story, as Adam and Eve are leaving the Garden of Eden in shame, having already disobeyed the injunction not to partake of the Tree of Knowledge, they pass another tree: the Tree of Life. God is so concerned that human beings not be in contact with the Tree of Life that an angel is called down to stand in front of the Tree of Life with a flashing, flaming sword, to forever guard it from human intervention.[3]

The meaning of the extraordinarily vivid imagery in this story tells us something important about biotechnological intervention: the Western world that invented recombined life is enjoined by its own tradition from interfering in the making of life. Only a higher power is supposed to be able to create life. No wonder many people in Western traditions have a visceral discomfort with intervention in the life-creation process.

We have lost sight, perhaps, of the meaning of the angel and her sword. The creators of the Genesis stories did not know about or anticipate biotech or recombinant DNA techniques. But they did know about human arrogance, greed, and short-sightedness. The lessons of ancient books like the Old Testament can be endlessly contemporary, providing people in the Western, science-oriented tradition with collections of values and precepts to learn from as we construct our moral lives in the modern epoch. No matter how literally or metaphorically we interpret these old tales, we ignore the swords of angels at our peril.

The venerable rules about diet in the ancient religious traditions also speak quite directly to genetically engineered food. The dietary restrictions in Old Testament religions restrict people from eating certain foods, or certain foods at special times. More to the point in our concerns about GM food, these traditions also forbid people from combining some natural items: milk may not be combined with meat, nor grains with grapes, not even linen with wool. There is a rich tradition of scholarship in interpreting these rules against biological combination,[4] which portray God as "the pre-eminent gardener."[5] The overall lesson is that in mixing foods we are mixing life and death, crossing life boundaries, and trespassing as in Genesis onto the territory of life creation that does not welcome human interference.

One does not need to be a member of a Western religion, or any religion, to experience a kind of spiritual or moral discomfort with the engineering of our food. There is a philosophical division in our early twenty-first century world between those who see themselves as a part of nature and those who see themselves as standing outside of nature, in a special category, not subject to the rules or processes that seem to govern life on our world.

It is not an accident that the altering of life at the genetic level is called "engineering." In the process of engineering, one begins with the end: a picture of the desired outcome. In fact, the engineering process starts with a design, is governed by the process of efficiency, and ends with the goal of production. Contrast this with what might be called the ecological approach, which begins with discernment – seeing what is there. Ecology is governed by the process of harmony and has no end goal at all.

APPROACH	STARTS WITH	USES PROCESS OF	GOAL
Engineering	Design	Efficiency	Production
Ecology	Discernment	Harmony	Harmony
Religion	Faith	Understanding and tradition	Salvation and enlightenment

One might wonder what is wrong with engineering food. The problem is that all human processes, including genetic engineering, have values embedded in them. If we substitute the values of design, efficiency, and production for discernment and harmony, what changes? The answer is, quite a lot. In the ecological approach to food, activity is oriented toward understanding what is there: discerning the processes by which plants live in the world and provide us with food. The values inside the idea of harmony incorporate notions of balance, integration, intrinsic function – and humility.

In engineering, by orienting ourselves toward a production goal we ignore the checks and balances of nature. We try to increase production (crop yield and monetary return) by inserting genes from other living organisms in ways that could never occur in nature, thus bypassing the naturally occurring boundaries nature provides us. Genetic engineers believe that they can "control the variables," that is, deal with the consequences of redefining nature's boundaries. They somehow always seem surprised when there are "unanticipated consequences" – when butterflies die, people develop antibiotic resistance, genes jump to nearby organic plants or weeds, insects develop resistance to pesticides, family farms close, people have allergic reactions, and the other items on the long list of consequences of using engineering values to dominate nature.

Actually, most of these "unanticipated" consequences could be easily anticipated if genetic engineers adopted an ecological set of values. But then, perhaps, they would no longer be genetic engineers – they would be ecologists, and they would not be inclined to engineer life to achieve their goals.

The genetic engineering of our food raises environmental, health, and social challenges, as we have seen. Engineering the life that provides our food also brings with it challenges to the moral and spiritual aspects of our long history as food growers and eaters, and as conscious beings. Before we accept genetically engineered food, we would do well to pay careful attention to the warning offered to us by Native Americans of the Six Nations Confederacy almost twenty-five years ago, before such food had even been invented:

> The way of life known as Western Civilization is on a death path on which their own culture has no viable answers. When faced with the reality of their own destructiveness, they can only go forward into areas of more efficient destruction... When the last of the Natural Way of Life is gone, all hope for human survival will be gone with it.[6]

We have the choice, as a people, to acquiesce to the unnecessary, unwanted, and hazardous path offered by the engineers of life, or we can follow our own deep traditions and values to choose an abundant, safe, and healthy food future for our families, our communities, and our nations.

8
WHAT THE FUTURE HOLDS

It's not an accident that many well-known science-fiction writers are scientists in their day jobs. The difficulty for us may arise when scientists lose sight of the line between speculation and fact. Even worse is when journalists and politicians start confusing hyperbole meant for investors with the truth. The rest of us remember the long parade of twentieth-century exaggerations, from the unsinkable Titanic to nuclear power too cheap to meter to genetically engineered crops that would feed the world. Science fiction is fine to read, as long as we keep the fabrications and fables on the bookshelves and out of our fields.

Yet we ignore the claims of GM food companies at our peril, because some piece of what they predict could come to pass. In this chapter we are going to sort some fact from fiction and gaze with the GM food fans into their crystal ball. After a tour of coming afflictions, we'll briefly look at some of the reasons to be cautious about biotech's brave new world.

Changing the Nature of Nature

FARMACEUTICALS

Plants as delivery systems for vaccines and pharmaceuticals is where we are. We can't get blood from turnips, but we'll be able to get blood protein in potatoes.[1]

Arnold Foudin, Animal and Plant Health Inspection Service

In our tour of some of the surprises that genetic engineers may have in store for us, the first is "farmaceuticals," or plants that produce antibodies, vaccines, pharmaceuticals, and human proteins. The Monsanto corporation has an integrated protein technology (IPT) unit to produce transgenic pharmaceutical proteins, vaccines, and industrial enzymes in plants.[2] Along with business partner NeoRx, Monsanto has genetically engineered corn plants that express monoclonal antibodies, and they have advanced to phase 1 and 2 clinical trials. According to William White, business leader of Monsanto's IPT unit, "If we grow [a pharmaceutical] in an acre of corn, or tobacco, the actual cost of production for crude product is negligible."[3] In this system, the plants are considered to be "factories" doing the work that was previously done by technicians in a laboratory at a much lower capital expense. The plants are engineered to produce the "crude" pharmaceutical product, which is then purified by the pharmaceutical company, to be packaged and sold to the public.

As part of his job as Assistant Secretary of USDA Animal and Plant Health Inspection Service, Arnold Foudin reviews the applications that agribusiness giants file, detailing their future projects and commercialisation technologies. Some of the developments that the USDA currently has under review or has already approved include rabies-vaccine corn, blood-protein potatoes, and diarrhoea-vaccine bananas.[4] In Canada, genetic engineers are working on blueberries that produce high levels of antioxidants and cranberries "designed to cure urinary tract infections."[5] Foudin believes that "corporations worldwide have seen the light," as companies like DuPont learn that "in the future your field will be the synthetic chemical plant of the world."[6]

Dutch researchers are taking a different approach by breeding genetically engineered plants that can produce drugs and vaccines in nectar, to be eaten in the form of honey. The bees that turn the nectar into honey will, supposedly, be kept in greenhouses so that they cannot escape into the wild or feed on any unmodified plants.[7]

Promar International, a firm that conducts studies of food issues, published a report, "Farmaceuticals and Pharming," available for purchase at the ambitious price of $15,000. Promar International envisions a bright future for farmaceuticals, suggesting that "oranges that replace daily multivitamins, potatoes with higher starch content that will not absorb fat, and soybeans containing insulin for the treatment of diabetes" are just some of the possibilities.[8]

In the summer of 1999 the first field test of plants containing human genes was conducted in Canada. The test involved tobacco plants engineered with the human gene that produces interleukin-10, a protein that acts as an anti-inflammatory in humans.[9] As they develop, the tobacco plants will also produce interleukin-10 in their tissues, and the hope is that the protein can be inexpensively harvested, isolated, and used on people. It will be several years before conclusive results are in.

MEDICAL FOODS: VALUE ADDED?

If genetic engineers can't deliver new drugs and vaccines in plants, then they are certainly hoping to deliver vitamins and minerals. These new foods are often called "medical foods," "nutraceuticals," or "functional foods." These foods are different from foods that have been fortified, or are naturally nutritious, because they are genetically engineered to contain very high levels of vitamins or minerals. Monsanto's highbeta-carotene oil, announced in March 1999, could be used to increase the level of vitamin A in humans who eat it.[10] Other examples include peanuts with better protein complementarity (from corn genes), higher-starch potatoes that will absorb less oil when fried, garlic that produces more allicin (believed to help lower cholesterol), and rice with higher protein (from pea genes).[11] Industry experts expect the nutraceutical/functional food market to go from $25 billion to $100 billion in the next three to five years.[12]

Genetic engineers hope that these foods will "work," but whether or not they will actually lower cholesterol, boost nutrient levels, or solve vitamin deficiencies remains to be seen. A good deal of hype surrounds the development of genetically engineered functional food crops, and as we'll see at the end of this chapter, their development raises far more questions than answers. What is clear, however, is that the kinds of failures we have already seen with genetically engineered crops could be even worse in crops having such a tremendous impact on our health.

For instance, many women need iron as part of a healthy diet. Women's bodies can handle iron supplements because they expel extra iron during their menstrual cycle. Men have much less ability to rid their bodies of excess iron, so vitamin manufacturers often leave iron out of men's nutritional supplements. If iron levels are too high in the human body, it can cause neurological damage. How will genetically engineered "medical" crops be segregated? How can the public be assured of protection against vitamin overdose? Will people need to cook two meals, one for women and one for men – and perhaps a third for the kids? Too much of certain vitamins and minerals can be just as damaging as vitamin and mineral deficiencies. Following in the monopolistic agribusiness tradition, these new crops would be corporate property. Writing in 1999, *Genetic*

Engineering News reported, "The Nutritional Research and Education Act (NREA), proposed by Sen. Orrin Hatch in 1989 and championed this year by Dr. DeFelice, is expected to be re-introduced into the US Congress this spring. The bill is based on the Orphan Drug Act and allows any company that conducts research on nutraceuticals to have an exclusive claim based on that research for seven years."[13] This means that nutritional advances will likely be controlled by a small handful of agribusiness corporations.

TAKE A POTATO AND CALL ME IN THE MORNING
Rather than a trip to the doctor to get vaccinated, you may soon have to go no further than the nearest farmer's field. Plants are being genetically engineered with human viruses so that when they are eaten, they become a form of oral vaccine. Bioengineers hope that by engineering a plant with genes linked to a human disease, the plant's leaves, stem, and other tissue will develop antigenic proteins that trigger antibody development in humans. The journal *Nature Medicine* reported in May 1998 that people who ate potatoes genetically engineered with a strain of *E. coli* that normally causes diarrhoea produced specific antibody-secreting cells that protected them from symptoms normally caused by *E. coli*.[14] This experiment demonstrated that transgenic potatoes could have an effect on the immune system, thereby setting a powerful precedent in the world of plant engineering and human health.

In other experiments, scientists inserted a non-toxic cholera gene into a potato, which when eaten triggered the body's production of antibodies against cholera. Even when the potato was cooked, half of the vaccine survived in live form. The scientists suggest that "one cooked potato a week for a month should provide enough active B-protein to immunise against the cholera toxin. However, because immunity falls over time, periodic booster spuds would be required."[15]

But you won't be able to grow these potatoes in your backyard. Potatoes like these, and many other new transgenic foods, would likely be considered "medical foods" by the FDA – given by a physician and intended as treatment for a dietary or medical condition. It is unclear where they would be grown, if they would be grown in confinement, how they would be processed, or to whom they would be sold.

An even greater problem with pharmacy foods is that in many ways they are a significant step backward. Pure drugs (the kind we take all the time in pill form), when not engineered into food, are more transportable, less perishable, less likely to hasten development of resistant organisms, and need only be taken in the exact amounts needed.

HEAVY METALS

During the discussion of proposed organic standards in 1998, three "hot" issues emerged: genetic engineering, the use of industrial sludge, and food irradiation. All three were left out of the revised organic standards due to overwhelming consumer objection. The issue of what to do with industrial waste has not been resolved, so genetic engineering is being looked to for solutions. The problem that needs solving involves agricultural land in the United States and the rest of the world that has suffered from the dumping of toxic chemicals, industrial waste, and poisonous heavy metals. Areas dumped on by large corporations have given rise to countless stories of polluted drinking water, horrendous health effects on humans and animals, and barren wastelands.

It is a great irony that agribusiness now wants to *solve* the problem of toxic soil, because many of these companies played a major role in creating the problem they plan to get rich fixing. Heavy metals can build up in soil that has been contaminated by industry or heavy fertiliser application.[16]

We know that many corporations now involved in genetic engineering were former chemical companies, or have agrochemical divisions. Monsanto and these other companies have toxic legacies that they are trying to leave behind.

Plant geneticists have been experimenting with plants genetically engineered to withstand high levels of heavy metals and industrial waste. These metal-tolerant plants can grow in soil conditions that would kill normal plants. They would supposedly absorb the metals and store them in their roots, leaves, and stems.

It is unclear at this stage how these plants will be disposed of once they have absorbed metals in the soil. Don't they constitute a toxic-waste problem of their own? How these plants will behave in the ecosystem has not been explained, nor whether they could pass on their heavy-metal absorbing properties to other food plants. What will the effect of the concentrated waste be on foraging animals?

INCREASING RATES OF ANTIBIOTIC RESISTANCE

The future direction of plant genetic engineering follows from the continuation of present trends. For instance, the continued use of antibiotic marker genes, which confer antibiotic resistance in genetically engineered food crops, makes the growth of antibiotic-resistant bacteria inevitable. According to the US Center for Disease Control and Prevention, there has been a significant rise in the appearance of antibiotic-resistant bacteria in humans. For example, the percentage of salmonella bacteria that resisted five commonly used antibiotics was negligible in the 1980s. By 1995 the percentage had risen to approximately 20 percent, and by 1997, 36 percent.[17] These dramatic increases show that the antibiotics used in our food products (both livestock and crops) are having a serious impact on human health. As long as genetic engineers use antibiotic resistance genes to create mutant food crops, human health and our ability to fight disease could continue to deteriorate.

GENETIC POLLUTION
We simply do not know the long-term consequences for human health and the wider environment [of genetically modified crops]...If something does go badly wrong, we will be faced with the problem of clearing up a kind of pollution which is self-perpetuating. I am not convinced that anyone has the first idea of how this could be done."[18]

Charles, Prince of Wales

As genetically modified organisms proliferate into unintended hosts, the problem of "bio-pollution" becomes more of a reality. The British courts have already heard cases regarding genetic pollution. In February 1999, Monsanto was fined $25,000 for having inadequate barriers between its fields of genetically modified crops and neighbouring fields of natural crops.[19]

Benny Haerlin, of Greenpeace International, says, "One way to estimate the environmental impact of GMOs is to look at the introduction of so-called 'exotic species,' i.e., plants or animals introduced into a new environment by humans or human activities. The most famous example is probably the introduction of rabbits to Australia. A rough rule says that out of 1000 exotic species, 100 will spread into the new environment, 10 will establish there and one will become a pest. The United States Agricultural Department recently estimated that the losses caused by these pest species amount to $123 billion per year in the US alone. And the international Union for the Conservation of Nature names the introduction of exotic species as a prime cause for extinction of other species."[20]

The Greenpeace statement brings to mind the case of the gypsy moth, introduced from Europe. Without any natural predators in the United States, the pest continues to ravage hardwood forests of north-eastern America. Who knows which natural species might be driven toward extinction by competition with escaped genetically modified organisms?

WHAT'S NEW DOWN ON THE PHARM?
The potential profit margins for plants that produce medicines or vitamins or that eat toxic waste could be exceptional, even by biotech standards. Perhaps the glint of gold on the horizon has blinded the would-be pharmers to the long list of problems that could accompany this kind of technology. Many of the potential difficulties are familiar to us from first-generation GM food. As we have seen, the transfer of genes from one organism to another is a chancy and imprecise affair. What other genes might be transferred, and then combine with medicines, or unknown toxic waste? What kind of substances might be produced once the foreign genes are in a new and somewhat variable environment?

The compilation of dangers goes on. What could happen in our environment when drug-expressing genes escape – transfer to other plants and animals? What could happen to the complex life in the soil when drug or vitamin or vaccine plants are ploughed under? And what about people who for reasons of allergy or personal belief do not want to take the drugs in the food, but the genes have jumped to common, unavoidable, and unknown food plants? As for the Dutch fellow mentioned above who is planning to keep his genbees in a greenhouse, we need only think of the so-called killer bees that escaped from captivity in Brazil. These introduced bees have spread all the way up to the southern United States and are still moving.

These problems seem familiar because they are: the basic critique of genetic engineering is the same for "generation two" biotech as it has been for the present round. Genetic engineering is an unasked-for technology thrust upon us by those who would make profits from it. It is an unproven and poorly tested technology, it is a technology dependent on new and inadequately controlled techniques, and it is a technology based on the release of organisms into the environment whose aggressive but dimly understood reproduction threatens the entire ecosystem.

Recently we were invited to speak to a college class. This is a common request, but in this instance eyebrows went up when the invitation turned out to come from a class in science-fiction writing. The good news is that many of the science-fiction fantasies of biotechnology's proponents have not come to pass, and many can't. With diligence and the kind of concerted action covered in the next chapter, we can keep fact separated from fiction, to the betterment of our world.

9
THE LIGHT AT THE END OF THE TUNNEL

What You Can Do

The wide spread of GM food, especially in the United States, is frightening. With upwards of seventy million acres of genetically engineered crops under cultivation worldwide by 1998, it may seem like our individual- and community-based decisions have little impact. Yet a lesson we can learn from the great Nestle boycott of the 1970s is that the very size of the problem can be an advantage for those of us seeking to make change. During the boycott, Nestle was the most ubiquitous food corporation on the planet, with yearly sales in the hundreds of millions. Activists sought to stop Nestle's infant-formula sales practices in poor countries that resulted in the death of babies.

The boycott campaigners realised that, because Nestle was everywhere, they could organise everywhere; the activists could create a worldwide network of sister organisations and supporters so that no matter where Nestle turned, the corporation would find opposition. The advantages enjoyed by transnational corporations – the ability to create global strategies, coordinate information, and share resources – are also available to those of us who want to stop the proliferation of GM food.

Twenty-five years ago the Nestle activists created an effective international activist network in sixty-six

123

countries. Just think what we can do in this era of faxes and instant internet communication. In fact, as we will see in this chapter, people in the United States and Europe have already won some significant victories over genetically engineered food.

In this chapter we are going to explore the specifics of what we can do in our daily lives, in our communities and nations, and on an international scale to preserve the safety and integrity of our food supply. Following this chapter are appendices of resources to help in the linking-up process.

STRATEGY ONE: BUY CERTIFIED ORGANIC
Unlike many earlier social movements, genetically engineered food touches every person's life immediately and directly. Each person is therefore provided with daily opportunities to engage in quiet, personal activism to reshape our food supply. The most straightforward method of keeping GM food out of our bodies is to keep it out of our kitchens. In many locales this is possible. In the United States, Europe, and much of the rest of the world, people can still buy food produced without the heavy use of added chemicals – food that was grown using more sustainable, earth-friendly culture. In the United States this food is labelled "organic," and those growers who have been certified by a third party adhere to the highest standards of food production. In some countries this kind of food is labelled "biologic."

Organic food is not GM food, as long as the farmers plant their crops from non bio-engineered seed. By taking the extra effort often required to find organic food, we can be assured that we are buying food produced without the pitfalls of genetically engineered crops. In all areas, *fresh*, non-processed organic food is free of genetically modified organisms. Unfortunately, some processed organic food, especially if it contains soy, might be contaminated with genetically engineered ingredients. Later in this chapter we'll look at some of the ways we can deal with these more subtle and complicated issues. But in general, if the label says "organic" or whatever term is used locally for this kind of sustainable, chemical-free growing, chances are the food does not contain genetically engineered ingredients.

Organic food is a great bargain. While the shelf price is often more than similar-appearing food grown in the chemical-intensive method, many hidden bonuses come along with the purchase price. The method of growing organic food builds our precious soil instead of depleting it, results in clear instead of polluted run-off water, employs more people in small-scale farming, preserves the health of farm workers and their children, contributes to the health of the people who eat the food instead of threatening that health with dangerous synthetic chemicals, and helps keep alive the traditional seed varieties that are the foundation of all of humanity's food supply. To receive all of this extra benefit for a few pennies more is one of the great bargains of the century!

In buying this kind of food, it is important to be aware of label tricks. Food sold with labels that say things like "natural" or "grown without additional chemicals" are not organic. Not only is there no guarantee that vaguely labelled food does not contain genetically engineered crops, but there is no standard at all to such made-up labels. Farmers who go through the trouble and expense of obtaining and maintaining organic certification should be able to count on informed customers to repay their commitment to safe food, and those of us who buy the food need to be able to rely on the label. It is important to find the label in your locale that means organic and buy that kind of food to the greatest extent possible.

For many people it is not practical to buy only organic food. As we saw in chapter 4, this is where labelling of GM food comes in. For people in some Western European countries where genetically engineered food is clearly labelled, they can simply choose food that does not bear the GM food label. In the United States, we must continue to push for adequate labelling of genetically engineered food.

Where in the United States can we find organic food? Some communities, especially in urban areas and in college towns, have food coops or buying clubs. Sometimes elaborate, with paid staff, and sometimes just a few neighbours sorting bulk-purchased food in a church basement, food coops can be an inexpensive and enjoyable way to obtain organic food. More and more communities are also holding farmer's markets, frequently just an area of a park or even a parking lot where growers can gather on an appointed day to sell their produce directly to the people who will eat the food. There is nothing quite like finding out precisely what kind of seeds and what style of farming went into growing your food, straight from the mouth of the person who grew it!

Similarly, increasing numbers of communities offer some version of a food plan called a "CSA," which stands for community-supported agriculture. Farmers who work near cities or suburban areas sell subscriptions to their produce. This enables them to raise capital at the start of the season and to share the risk of agriculture with the community that benefits from their work. In exchange for sharing the risk, the subscribers receive generous weekly boxes of very fresh and healthy vegetables, and sometimes fruits, herbs, and even flowers, right from the farm. It is important, when joining a CSA, to get a clear understanding of the group's expectations as well as just what sort of agriculture the farmer is using. In most cases, the boxes are packed each week with GMO-free, delicious food.

Increasing numbers of urban areas in the United States and elsewhere have natural food markets, some of them quite large. As long as the shopper is very careful to read labels, since some such markets mix organic with chemical-intensive food, a good variety of both fresh and preserved healthy food can be obtained.

Finally, if all else fails, it is worth investigating mainstream supermarkets. American supermarkets run on paper-thin profit margins, usually around 1 percent. The supermarkets are therefore always on the lookout for ways to bring people into their stores. These supermarkets almost universally use laser scanners to read the prices of the food as shoppers check out. Those scanners do a lot more than tell you the price of your breakfast cereal: they tell the managers of the supermarket exactly how much of each item is being sold. This makes supermarkets ideal for "reverse engineering" by shoppers who vote daily at the cash register. More and more supermarkets are offering organic items. If a few organic tomatoes or some frozen organic peas sell well, the company will rapidly order more. Similarly, as genetically engineered foods start to be labelled and the supermarkets find the food labelled GMO-free selling well, they will place their orders accordingly. Even if it means an extra ten minutes of travel, going out of our way to give our business to merchants who take a pledge to sell natural, non-engineered food has a double-bonus effect. It ensures that we eat the food we think is best for us and our families, while at the same time rewarding those who take the risk of announcing a policy to sell only healthy, natural food.

Big multinational companies support the free market. Fine – let's make that market our tool, too, by pushing our financial clout in the right direction. When we can buy food that is not genetically engineered, even if it means a bit more trouble or cost, we are using our purchasing power to help reshape what is grown.

STRATEGY TWO: GROW YOUR OWN
Gardening is a favourite pastime around the world. While few of us grow all of our food, and many people live in cities or in climates where the amount of food they can grow is

limited, most people can grow something, even if it is some basil on a kitchen window ledge or a pot of tomatoes on a fire escape.

As visionary gardener John Jeavons has shown us, even a very small plot of land can grow a lot of food if properly cultivated. While most of us cannot feed our families year round, we can come up with bushels of wondrous fresh tomatoes, piquant spices, and unusual salad additions. An especially useful strategy is to see what foods are not available locally in organic form that you would especially miss at your table.

This can be the priority for your garden. Don't leave out fruits. Some fruit trees come in dwarf varieties and require surprisingly little maintenance. Some of the older heirloom varieties bring the added bonus of producing fruit gradually, so one tree can supply a family for months, in contrast with the "modern, improved" trees that have been modified to ripen their fruit all at once to focus the picking season and thus reduce labour costs.

A second reason to garden is to educate our children. Far too many children think that food comes from supermarkets, wrapped in plastic. Children are the consumers and voters of tomorrow. The willingness of children to accept GM food and other factory versions of the natural world will be decided partly on their experience of the natural world during their early years. Children's memories of the smell of warm earth in the spring, and of the miracle of life that they experience when they gently push tiny seeds into the earth and weeks later taste the sweetness of fresh-picked peas from their own vine, will stay with them for their entire lives.

A small, unambitious garden can be especially valuable recreation for a modern, busy family. Children can then focus on the joy of the work, learning a valuable lesson while eating healthy food produced by their own labour. Children who are able to do this might present a less

The Light At The End Of The Tunnel

receptive countenance when they are inevitably exposed to the smooth talk of big greedy GM food companies.

A third reason to garden as a response to genetically engineered food is to establish your own little seed bank. The struggle over food is ultimately a struggle over seeds, the miracle life packets that keep us all alive. As we have seen earlier in this book, GM food necessitates a kind of monoculture, the unnatural biological sameness that is so dangerous. Even in the United States, the world capital of GM food production, backyard gardeners who save seeds still far outnumber commercial seed sources. Yet the extinction rate for non-engineered food-producing seeds, as documented by the Iowa-based Seed Savers Exchange, is still rapid. In fact, the majority of food-producing plants that were available one hundred years ago are now extinct – gone without hope of recovery.

Since 1981 Seed Savers has been publishing a catalogue of catalogues, a listing of every vegetable seed variety offered commercially in the United States. The 1999 edition of the *Garden Seed Inventory* lists 7,313 vegetable varieties in over eight hundred pages, from Amaranth to Watermelon.[1] The good news is that while seed companies continue to consolidate or go out of business, thus endangering their seeds with extinction, recent years have shown a solid trend toward establishment of new seed companies and a consequent increase in the number of new vegetable varieties offered for sale.

Why would a backyard gardener care if a seed company goes out of business? There are over two hundred such companies, after all. The answer can be found in the statistics of the *Garden Seed Inventory*: more than 50 percent of all vegetable seed varieties offered for sale are only available from one company. If that company goes out of business or is absorbed by a larger, perhaps less-caring company, that variety can go extinct, fast. In the 1999 edition of the inventory, over one thousand varieties are shown as "dropped," meaning that some could already be gone forever.

If we save seeds that we grow, vegetable varieties dropped from single-source catalogues might thwart extinction in the old baby food jars and freezer packets of dedicated backyard gardeners. Caring for the seeds we grow is a way of ensuring that we will have a diverse, healthy, and plentiful alternative to genetically engineered seeds. Saving the seeds of most plants is simple. For most people, just saving seeds over the winter is all that is required. In doing so, you can save the seeds of those plants that grow best for your yard and your tastes. Over time you'll have your own subvariety, especially tailored to your needs. This is the time-honoured partnership between people and their food plants.

Because you will save seeds from plants that do well in your locale, you'll have the chance to thrill neighbouring gardeners with a few extra seeds that will do well for them, too. Aside from the reward of generosity, you'll know that in dispersing your seeds you are keeping them safe: if the dog eats your seed supply or you forget to water during a heat wave and your plants die, the seeds you gave to your neighbour might have survived, so you can ask for a few back at the end of the season. That's really all a seed bank is.

If you get especially ambitious or come up with some special plant that you want to share – and preserve – you can contact a regional seed saving program or join a national seed-savers organisation. This will enable you to not only share your bounty, but to then become eligible to see what other backyard gardeners have grown for sharing. See the appendices for a list of some of the growing number of such organisations, as well as a guide to seed saving.

STRATEGY THREE: THE POWER OF THE CALENDAR
As we have seen in this book, genetically engineered food is

part of the process of making our food supply a global profit centre for a few corporations. In fact, it can be argued that GM foods don't make sense unless they are marketed on the global stage. Yet as we have also learned, our food supply is safest when it is localised, adapted to the conditions of our varied world.

The process of globalisation in our food supply harms local people. Farmers in countries where food is imported find their prices driven down by competing food flown in from places where the economics are entirely different. Some farmers just go out of business. In the countries that become food exporters, the local farmers become contract labourers, forced to plant alien seeds – even if local food varieties are in danger of extinction as a result – and obliged to spray the chemicals called for in their contracts. The result can be and has been a displacement of small farmers in both sending and receiving countries and drastic changes in the availability of traditional foodstuffs.

Export food is no great bargain for shoppers, either. Those who import the food are looking for the best price on the local market. They might be a lot less concerned with issues like the amount of pesticides and other chemicals sprayed on the food, the health and safety of farm workers and their families, and, of course, whether or not the seeds have been genetically modified.

In some places, local custom or even regulation causes food markets to post the origin of the food that they sell. Unfortunately, in the United States it is difficult if not impossible to determine the origin of most fresh food. When origin can be determined, it is best to buy as locally as possible. Best of all is a visit to a farmer's market or roadside stand where the farmer herself can be questioned about the use of genetically engineered seeds and chemical use.

When the origin cannot be ascertained, we still have our secret tool: the calendar. Almost all the food we eat ripens according to the season: peas in summer, squash in the fall. The food industry beats this system by taking advantage of the earth's equator: when it is winter thirty degrees north of the equator, it is summer thirty degrees south. Because the great central valley of Chile is a geographic mirror of the San Joaquin Valley of California, which produces more of the fresh produce in the United States than any other location, Chile has become a major food exporter – at considerable cost to indigenous agriculture in that country.

While it is nice to be able to buy asparagus any time of year, we are supporting the globalised food industry most of all when we insist on ignoring the natural calendar of food. Food is a product of biology, which is tied to the rhythms and cycles of life. Unlike toasters or books, fruits and vegetables can't just be plucked off the shelf whenever we choose to make our purchase. Because so many of us have become used to the cornucopia displays in modern supermarkets, we have forgotten much of what our parents and grandparents knew about the seasonality of fresh food.

Buying food according to the natural calendar will not solve the problem of genetically engineered food, but it provides a counterforce to the globalised food corporations that depend on GM food. Buying in season also strengthens local farming, not just in your local community but around the world. As we have seen, it is within local communities that sustainable agriculture decisions are best made, increasing our own chances for healthy and safe food.

STRATEGY FOUR: LINK UP
We may have new life sciences, but we have new citizens groups too. For many of us, simply buying organic food and maybe growing a few vegetables is all we have the time and inclination for. Increasing numbers of people, however, alarmed at this unasked-for change in the food that sustains them, want to join with others to make changes in the larger arena. In the next section, we'll look at some of the

activities that are available. But for those who want to become more active, as well as those who don't, the place to start is with more information. A characteristic of the struggle over GM food is that wealthy corporations can purchase constant barrages of "public relations" information – and disinformation – in the media. To counter this we need to exert just a bit of energy to gain unbiased information, free from conflicts of interest, oriented to the public good.

Luckily, reliable information about genetically engineered food abounds, from magazines like *GeneWatch* and *The Ecologist* to a huge number of sites on the World Wide Web. Even if you don't feel interested in joining a group, even if you don't have the time or the inclination for much investigation, a connection to one or two reliable sources of information can be an antidote to the self-serving and biased propaganda from corporate agribusiness.

In putting together the appendices in this book we have tried to provide you with a rich diversity of sources of information in books, magazines, web sites, and other resources. Because the subject of genetically engineered food is growing and expanding, we encourage you to keep monitoring what is available. With a minimal investment of time and money, a person can become well informed, and information is a powerful tool in the conflict over our food supply. Truth is strong and truth is cheap.

STRATEGY FIVE: ACTIVATE!

It is natural for many of us who become informed about the problems of GM food that have been described in this book to want to *do something!* Just as nature thrives on diversity, so does social change and activism. There is a rich variety of possible actions that a person can take to address the genetic engineering issue.

One of the best and most enjoyable forms of activism is

to join a local organisation. Local groups can be the most democratic, they operate appropriately within our local culture, and they are usually small enough that a person can quickly feel the difference she or he makes. With the exception of some parts of Western Europe and a few other locales, there are not that many community-based organisations addressing the problem of genetically engineered food. However, few communities are lacking in a local ecology centre, environmental group, organic farming or gardening group, or similar organisation. While the basic purpose of local groups and clubs should be respected, many will welcome an appropriate tabling of new issues that affect their concerns. (And if an appropriate group doesn't already exist, start one!) Studying and learning about a new issue as a group can be enjoyable and more productive than solitary investigation.

The great strength of local organisations is that you can work to modify local institutions. What are the rules for purchasing food served in the local school? How will the manager of a local market react when visited by a friendly delegation of local citizens, asking for hand-lettered signs showing country of origin? Local newspapers will be more likely to cover even a small public meeting, leading to further public education. Local groups can visit – or foster – farmer's markets to see what is sold, what is labelled, and how. You can also encourage local restaurants to buy local produce and organic foods – a growing number will even label their menus. Some chefs and restaurants have taken a pledge to use local, organic food as much as possible. Look for this information in the restaurants – or ask.

Each member of a local group becomes an agent for change elsewhere. One member might find herself on a committee for a church dinner, where she can raise the issue of GM food and perhaps share copies of a simple pamphlet her local food group has drafted. Another group member can serve a non GM food snack at her child's class, while the entire group takes turns attending the regional coordinating group sessions once a month.

An excellent addition to local activism is a tangible connection with a national organisation. Groups like our Council for Responsible Genetics publish periodic newsletters that can furnish a local activist with clearly written expert material at a nominal cost. In addition to receiving valuable information, supporters of national organising will know that they are helping to maintain a research – and social – change presence that cannot easily obtain funds otherwise.

National groups like the Friends of the Earth or the Consumers Choice Council play an important role in addressing the genetic engineering revolution. These groups conduct objective research into the frequent dubious claims of the GM food companies. They maintain expert staff and board members who appear on panels, brief the media, and sometimes act as a resource to government and regulatory officials and staff.

This brings up the complicated subject of rules, regulations, and governments. While the content of school lunches can best be addressed at the local level, questions of government regulation and labels are often the territory of state or provincial governments, and sometimes federal or even international regulatory bodies. Petitions and letters from local groups can have a real effect on legislators, but ultimately individual citizens cannot count on having the political power to match the clout of companies like Monsanto, which as we have seen puts great resources into telling its story-including to the government.

There are at least two routes to dealing with the seemingly great power and influence of the GM food companies. One is to engage in citizen activism at the local level. For those who might be cynical about the value of sending letters and e-mails to government officials, we have the shining example of the immense outpouring of citizen concern over the proposed inclusion of genetically engineered food under the organic label, which caused the government to change their policy. In addition, there is the willingness of governments in countries like Great Britain to resist their cousins across the Atlantic and permit labelling of GM food.

Writing letters, calling local officials, holding public meetings, circulating pamphlets, and the other paraphernalia of citizen activism can have an effect, especially when it is part of a large-scale trend. The other route to changing government minds and policies is to work via the national organisations that are established partly just to collect citizen sentiment and deliver it in a sharply pointed fashion to just the right bureau or official.

National activist and some environmental groups are increasingly learning how to translate local sentiment about genetically engineered food in a way that creates change. This process has been more successful in some European countries than in the United States, but there is little doubt that as citizen activism grows, national groups will see an increase in their clout as well.

Creating change in corporate policy doesn't have to be difficult. In May 1999 Greenpeace sent a fax to the Gerber baby-food company inquiring about its use of genetically engineered ingredients. Within weeks, Gerber had changed suppliers so that its baby food would be free of such ingredients. What makes this impressive switch even more notable is that Gerber is owned by Novartis, the Swiss multinational that is the third-ranked seller of genetically engineered seeds in the United States.[2]

Further proof of the powerful effect activists have had on this issue came in September 1999, when Archer Daniels Midland, the giant buyer and exporter of farm commodities, requested that its suppliers segregate their GM food and conventional crops, because of its difficulty in selling the GM food. Planting of genetically engineered corn, initially projected to increase by about 20 percent in 2000, is now projected to drop by 20 to 25 percent.[3] That means that there will be an additional 8.3 million acres of corn grown in the United States in the year 2000 that isn't toxic to monarch butterflies.[4]

When one has joined a local group, sent a contribution to a reputable national non-governmental organisation, and written letters, what is next? For a small but highly significant number of people, there is direct action. This is a controversial area of activism, but one that is worth some examination. Direct action about genetically modified food has taken a number of forms. Some activists have publicly dumped milk containing bovine growth hormone. Especially in Great Britain, activists have trespassed on land where genetically engineered plants were undergoing field trials, pulling the plants up and leaving them in biohazard bags for proper disposal. In California, the Biotech Baking Brigade has launched tofu pies into the faces of biotech corporation officials, including Monsanto's Robert Shapiro.

Some people are uneasy with these tactics because they may involve breaking the law, they seem extreme, and the press coverage is sometimes negative, but those committed to direct action reply that in the tradition of Henry David Thoreau or Gandhi, some unjust laws need to be broken. They say that the real extreme crimes are those committed by officials of biotech companies who jeopardize public health and safety in the name of profit. And they point out that seemingly negative press still brings the matter to the attention of many more people than can be reached by leaflets or petition drives. The goal of these actions can be direct change, as in pulling up dangerous plants, or it can be simply dramatising a problem to get people to learn about it.

However one may feel about direct action, it is a part of many burgeoning social movements. Just as the natural world includes plants and animals that may frighten us – and may be inclined to sting us on occasion – a complete and healthy activist ecology must include true diversity in order to thrive. Some people, in their response to the challenge of GM food, will buy a few items with organic labels for the first time in their lives. Others will pull plants from experimental fields and end up in shackles before a magistrate. A vigorous democratic movement dealing with genetically engineered food will include these extremes and every permutation in between. The crucial point is that there be broad, diverse, and effective action by people concerned with GM food. The result of this concerted citizen action will be a healthy, safe, and varied food supply for all people.

DEALING WITH TOUGH CHOICES

Because GM food has invaded portions of our food supply so thoroughly, it is not possible to attempt to deal with the problem without encountering conflicts, puzzles, and even outright impossibilities. For example, some among the many millions of vegetarians rely on meatless burgers and other food that is made from soy. Given the skyrocketing percentage of soy that is genetically engineered and the lack of labelling in the United States and many other countries, a person seeking to substitute soy products like tofu and veggie burgers for meat will be faced with the dilemma of their favourite soy product containing genetically altered soybeans. One way to address this is to buy organic, but organic soy products are still not widely available. What's a conscientious person to do?

Another difficulty arises when two important values conflict. For example, there are increasing numbers of organic products available as imported food. While this is not much of a problem in the tightly integrated trade of Western Europe, it poses a problem when a person in a US market is faced with a choice of GM food from their own farms or organic food imported at high energy costs from a southern country that has decimated its indigenous agriculture in favour of the higher-profit export trade. Is it better to buy an organic bean from Chile or Mexico, a frozen or canned domestic organic bean, a fresh non-organic one, or none of the above? Other difficulties arise: the conflict people face when they want to keep a kosher or halal household, or what one can do when eating out in restaurants. While a small number of chefs have taken a pledge not to serve GM food, most have not and few restaurants label such considerations on their menus.

There are other instances of problems and conflicts in choosing a diet in these days of rapid change in how our food is produced and sold. It is important to recognise that few people will be able to maintain 100 percent GM food-free diets for very long, except perhaps those who grow all of their own food and live far enough from genetically engineered crops to avoid contamination from wind drift. This does not mean that genetic engineering of our food supply is here to stay. It does mean that we have to sometimes just do the best we can, and keep trying to do better.

Genetically engineered food appeared suddenly and it spread with startling rapidity. Companies like Monsanto and Novartis, with their global reach and vast resources, can appear invincible. Yet when we look at the response to this rash and greedy attempt to take over our food supply, we see a very encouraging picture. Millions of people have mobilised themselves into a diverse social movement in a matter of months. This increasingly global movement has had the power, in Europe and Japan, to move governments, get laws passed, thwart worldwide trade agreements, and sign up a variety of supporters from princes to school children. Even in the United States, birthplace of GM food technology, the sleeping giant of American public opinion is stirring from its slumber. Monsanto has much to fear.

Changing the Nature of Nature

Every generation or so, a social movement arises that defines that era, one that reminds people that masses of ordinary citizens have deep knowledge, extraordinary power, and legitimate moral authority. While the struggles over civil rights in the United States and over pollution worldwide continue, not yet won, we are well-advised to scorn the naysayers. Instead we can build the movement to eliminate genetically engineered food by remembering our victories – the fallen Berlin wall, the diminished arms race, the mothballed nuclear power plants. The hope felt by millions of people around the world that their children will live in a healthy, safe, and just world is a force more potent than all the quick profits of the GM food companies. With just a bit of attention from each of us, acting individually and together, we can restore a secure food supply that will nourish us now and for years to come.

POSTSCRIPT

BRITAIN IN GM REVOLT

By Peter Hounam

Public disquiet about biotechnologies caught fire in the UK in the late 1990s and, ironically, it may have been Tony Blair, the Prime Minister, who generated this backlash. As pressure groups and environmentalists loudly expressed their concern, and the media became increasingly sceptical, Blair appeared to have sided with the GM lobby.

Known as a supremely canny political operator, he was clearly dazzled by the commercial potential of GM. In February 1999 he played down the mounting concern, insisting the Government had taken the best scientific advice, and arguing that adequate precautions were being taken to avoid any problems. "There is no GM food that can be sold in this country without going through a very long regulatory process," he said on local radio. "Let's proceed on the basis of genuine scientific analysis and inquiry, proceed with very great care and caution and not get the facts mixed up."

Blair was being questioned as controversy mounted about the research of 68 year-old Arpad Pusztai. As mentioned in Chapter 3, he had fed rats with potatoes modified with lectins, a substance found in snowdrops known as GNA which protect plants from insects. The new super-spuds would also be resistant to attack, it was believed, but Pusztai claimed the potatoes caused many organs of rats enjoying them to differ in size compared with rats fed ordinary potatoes. Their immune system response was also reduced, he said, providing one of the first snippets of evidence that commercial GM techniques could be harmful.

Pusztai was working at the Rowett Research Institute near Aberdeen, where its director, Professor Philip James, surprised everyone by taking quick action to denigrate his senior colleague's results. The institute claimed to have checked the findings and had found no evidence of any such effects. Pusztai was suspended after appearing on a World in Action programme to explain his findings and then retired. As a close adviser to Mr Blair, James strongly denied ruining Pusztai's reputation, gagging him or acting from political motives. He said: "Arpad was not at all well during the episode and he was mightily relieved when I said I would handle the press inquiries at that time." The feeling was left, however, that an official GM cover-up was underway to play down problems and exaggerate the potential of GM research.

This feeling was strengthened in June 1999 when Blair was interviewed again, this time on BBC Television. He was not taking sides, he said, but it would be bad for British industry to give in to the public outcry over GM foods. Instead people should keep an open mind: "I just worry that we have to proceed according to basic evidence and not say because people talk about Frankenstein foods we [should] simply chuck the whole thing out of the window."

He warned against banning GM foods and biotechnology research and "then finding in 15 or 20 years that we'd got it wrong". "We are in the position as the Government where it is almost as if people say 'you are the great advocates of GM foods'. I'm not the advocate of anything other than keeping an open mind. All I say to people is just keep an open mind and let us proceed according to genuine

Postscript: Britain in GM Revolt

scientific evidence."

But Blair was clearly fired by the spirit of competition, and the desire for Britain not to be left behind. He said Germany "was pouring literally hundreds of millions of pounds" into biotechnology to catch up with the UK. "Genetic modification has many different areas, for example in medicine, and Britain is at the leading edge of this new technology...it could be the leading science of the 21st century."

Blair's wife Cherie was said to be much more sceptical, as was Michael Meacher, the Environment Minister. However, it later became clear that Blair was secretly under pressure from the US to prevent any restrictions to free trade in GM products. The Guardian newspaper revealed that prior to the Downing Street summit between Clinton and Blair in May 1998 the US President was carefully briefed to twist Blair's arm into supporting the interests of US GM companies.

Documents obtained under the US freedom of information act show that Clinton told Blair – who then held the EU Presidency – that "the EU's slow and non-transparent approval process for genetically modified organisms has cost US exporters hundreds of millions in lost sales." The paper went on: "In the spirit of increased US-EU regulatory co-operation, we urge the EU to take immediate action to ensure that these products receive a timely review."

The US's main worry was that Britain and the rest of the EU might insist on product manufacturers warning their customers whenever goods contained GM ingredients. The briefing paper concluded: "Differences among member (EU) states over labelling have been an impediment to reforming the approval process. We will be watching the commission's efforts to implement its new guidelines for labelling and we hope the EU can now move quickly to complete review of the products in the pipeline. That said, the US sees no reason to label a product simply because it has been genetically engineered. Mandatory labelling of GMOs should be based on sound science."

Blair seems to have reacted immediately. It was reported the next day that he had brought forward proposals to scrap EU plans, which had created an impasse with national governments, for labelling food that might contain GM ingredients. In line with US policy, Britain proposed to the rest of the EU that only US soya bean derivatives would have to be labelled.

Tony Juniper, a British Friends of the Earth campaigner on GM food was not surprised about the leak, though highly critical of Blair's apparent submissiveness to US pressure. "The government has repeatedly denied in private and public that the US has ever raised or tried to put pressure on the government over GM food. These previously secret briefing papers suggest the opposite occurred."

Blair was not alone among British politicians in decrying the doom-mongers and supporting a fast-paced research programme. In May 1999 Parliament's select committee on science and technology came out firmly in favour of taking some risks in the interests of further experiments. It said it was essential that information provided to the public was accurate and that debate was well-informed. "Labels such as 'Frankenstein' or 'Mutant' and sensational media reporting distort that debate and jeopardise rational policy making."

Its report concluded: "No human activity is risk-free. We have seen no evidence to suggest that the risks associated with growing GM crops or eating GM crops are high enough to justify calls that have been made for a moratorium; indeed we have seen no evidence to suggest that the risks associated with eating GM food are any higher than those associated with eating conventional food. There are widespread concerns about the impact of intensive farming practices on the environment and biodiversity. If properly managed GM technology can offer a solution to some of those problems although we accept that any such solutions will require responsible management of the technology..."

One major figure with a concern for the environment was not so sure. The Prince of Wales penned a newspaper article the following month which placed him firmly in the opposition camp. In the *Daily Mail* Prince Charles wrote: "The debate about the use of GM technology continues, with daily news of claims about the safety or the risks. The public's reaction shows instinctive nervousness about tampering with nature when we don't know all the consequences.

"There are unanswered questions which need to be asked – about the need for GM food, its safety, the environmental consequences, consumer choice and the usefulness to feed the world's growing population. At the end of last year I set up a discussion forum on my website on the question of GMOs. I wanted to encourage wider public debate about what I see as a fundamental issue and one which affects each and every one of us, and future generations. There was a huge response – some 10,000 replies have indicated that public concern about the use of GM technology has been growing. Many food producers and retailers have clearly felt the same overwhelming anxiety from their consumers who are demanding a choice in what they eat. A number of them have now banned GM ingredients from their own-brand products.

"But the debate continues to rage. Not a day goes by without some new piece of research claiming to demonstrate either the safety or the risks of GM technology. It is very hard for people to know just who is right. Few of us are able to interpret all the scientific information which is available – and even the experts don't always agree. But what I believe the public's reaction shows is that instinctively we are nervous about tampering with Nature when we can't be sure that we know enough about all the consequences."

Greenpeace had also become centrally involved in the issue, taking direct action to destroy GM crops in East Anglia. Lord Melchett, its head in the UK, was held in jail overnight for one highly publicised raid. For a former Labour minister in the House of Lords, it was an act of principled defiance that won support from a hitherto silent group of Labour backbenchers.

By late 1999 most of the major supermarket chains had banned products containing GM ingredients and officialdom was even more at odds with most of the national press in the UK with both the *Daily Mail* and *Daily Express* mounting campaigns against GM food. This sparked the ire of Sir Robert May, Blair's chief scientific adviser, head of the office of science and technology, as well as Royal Society research professor at the department of zoology, Oxford. He described coverage in the *Express* and *Mail* as "crap", provoking Rosie Boycott, editor of the *Express*, to accuse him of "vulgar contempt" for her paper and its readers. May was forced to weigh his words more carefully.

"What I said was quite well intentioned," he said, "(but) I now realise it was unfair to criticise the *Daily Mail* or the *Daily Express* because one ought to realise they are more into entertainment. There are many publics and I think it's a pity that the *Mail* and the *Express*, for complex reasons which are partly to do with commitment and partly circulation, have chosen to mount a campaign on an issue that is highly complicated. Of course, the moment you say that, you sound like another of those scientific idiots who say that everything is complicated."

May is a brilliant scientist and being an Australian he is not afraid to speak his mind about the dangers of overreaction. He told the *Guardian*: "The first crops to be trashed were frost-resistant strawberries in the late 1970s and early 1980s by people wearing decontamination suits – in America. There were calls for a moratorium on research in the States. Spearheading the movement was not so much Greenpeace as Science for the People – faculty groups on university campuses. At that time, it delayed the construction of the molecular biology building in Princeton. People were alarmed at the horrors that might be going on in laboratories at a time when there was very little fuss in this country. Now we have the exact opposite."

May admitted that the BSE disaster had heightened people's fears in Britain about scientific developments in agriculture. "The GM controversy is to do with recent events, and BSE is a particularly important one. But I believe the efficacy and sophistication of the lobby groups here has something to do with it. I also think it has to do with the sophistication of the newspapers in understanding what causes will engage their readers."

Perhaps May went back to Downing Street and began to urge a more cautious approach because by February 2000 Blair was beginning to seem much more open-minded. In an article in the *Independent on Sunday*, the Prime Minister said he understood the "cause for legitimate public concern" on the GM issue. "There's no doubt that there is potential for harm, both in terms of human safety and in the diversity of our environment, from GM foods and crops." Blair said his government was proceeding "very cautiously indeed" because of the potential ill effects.

Friends of the Earth immediately interpreted this as a volte face. Director Charles Secrett said: "At long last Mr Blair is listening to the public who have made it perfectly clear that they don't want GM food on their plates, or GM crops in their fields. Now it is essential that this approach is backed with action."

In fact Environment Minister Michael Meacher, long a sceptic, had already been working behind the scenes for greater controls of GM food imports and must also have had an impact on Blair's approach. The minister had attended an international gathering in Montreal a month earlier where 130 nations discussed GM issues. Overcoming opposition from the United States, Canada, Australia, Argentina, Uruguay and Chile, they adopted a protocol giving governments the right to block GM crop imports if there was "reasonable doubt" they could endanger public health or the environment. Previously firm evidence they could be harmful was required.

This "Biosafety Protocol" was said to have ended seven years of wrangling and campaigners said it was an important step forward. Meacher was credited with playing a leading role in forging the agreement. He said: "The financial markets are changing. The stock rating of Monsanto has fallen so much that Deutsche Bank has advised it to get out of biotechnology." He acknowledged that the Americans had wanted the minimum impact on free trade but the bulk of countries attending the conference had won "the right to say no".

Greenpeace commented: "This is an historic step towards protecting consumers and the environment from the dangers of genetic engineering. Common sense is starting to prevail. We are happy that the US and other opponents failed to force upon the world this untested and risky technology." Sarah Finch of the World Development Movement said EU ministers had played a crucial role in opposing the US-led faction: "The precautionary principle has been established after a long hard fight. There are more battles ahead but this an excellent start."

Despite the euphoria, those wanting a complete moratorium on GM research still had a long battle ahead. Greenpeace stepped up its direct-action activities and members boarded the 60,000-ton ship Iolcos Grace carrying genetically modified soya to Britain. Off the coast of Anglesey, North Wales, Greenpeace activists used an inflatable boat to reach the vessel and the six activists attached themselves to the anchor mechanism, preventing it from moving. Greenpeace then asked Cargill, the

Postscript: Britain in GM Revolt

American owners, to return the cargo to America.

Cargill Europe confirmed the Iolcos Grace contained GM and non-GM soya and that its main use was in feeding livestock for human consumption. Greenpeace said the agreement of the Biosafety Protocol in Montreal a few weeks earlier meant that governments including the UK could now refuse to accept imports of GM crops on the basis of the "precautionary principle". The protest ended – and the cargo was unloaded.

Faced with overwhelming public opposition to GM products from the British public, food industry experts now began advising supermarkets and processing companies how to obtain supplies of conventionally grown crops uncontaminated with GM ingredients. A deal was sought with soya and maize growers in the US who had resisted the trend to GM production.

The OECD now became centrally involved in the issue, hosting a conference in Edinburgh attended by senior politicians from all over the world including Britain's cabinet office minister, Mo Mowlam. She denied Blair had made a U-turn on the GM issue: "What we have always said is that there are potential harmful effects. There is always an element of risk. What is important in this issue is that the public have knowledge of the risks. We hope that by labelling and the research being done that that will be the case."

One of the speakers from India made a big impact. Surman Sahai, president of the Indian Gene Campaign, told delegates that biotechnology companies must ban so-called "terminator" crops. These crops cannot reproduce, she said, forcing farmers to buy new seed every year and thereby discriminating against third world farmers. She said GM firms like Monsanto had to end a culture of secrecy surrounding the use of their products in poorer countries. "One of the problems in India is that field tests were not transparent," she said. "The farmers in the region were not told what was happening...Those of us in the developing world cannot understand sterile technologies. If it is going to threaten our self-reliance in food production, if our self-reliance in agriculture is threatened, it is clear we will not accept these technologies."

Mowlam's attempt to reassure the world that Blair's New Labour Government had not changed sides disappointed leading British environment campaigner George Mombiot, a prolific writer on the GM issue. He said Mowlam's efforts were "sadly predictable"'. He went on: "Genetic engineering has always troubled Tony Blair, for it divides his two great constituencies – big business and Middle England. When faced, for the first time in his premiership, with the need to alienate one by supporting the other, he chose to side with the lobbyists, and against the electorate.

"It was only after the business case for GM food collapsed that he changed his position. Even so, while acknowledging the legitimacy of public concerns about the environment and food safety, his article on Sunday avoided the real issue: the corporate capture of the food chain. Mr Blair's backing for the biotech companies has, as he now appears to have recognised, done him untold harm. It has taught businesses that he's a dupe who will sing to their tune, however off-key it may be. It has encouraged voters to explore the other insalubrious liaisons in which he has become entangled. His conversion to neutrality, even in the narrowest political terms, is long overdue. But it will take a lot more than an article in a Sunday newspaper to make it look convincing. To recover his credibility, he must snatch government policy out of the of the biotech business."

As Mombiot realised, the battle was far from over. Full-scale trials of three GM crops across England and Wales were given the go-ahead in March by an independent committee of scientists. It confounded the hope of campaigners that too few farmers would be found who were willing to plant the crops. It was reported that ministers

Postscript: Britain in GM Revolt

had "put a rocket" under the biotech industry. The scientists then agreed that 15 fields of genetically modified oil seed rape, 16 fields of GM maize and 25 fields of mixed GM sugar and fodder beet would be planted and monitored regularly to check effects on animals and birds, weeds, and the spread of pollen. One major concern was whether they contaminated other crops or wild plants. No officials involved were allowed to comment but a terse statement was issued: "The UK has led the world by setting up these scientific trials," it said. "They will provide the important answers to key environmental questions. Ministers have made it perfectly clear that there will not be commercial growing in this country until they are satisfied there will be no unacceptable affects on the environment."

Greenpeace said it was "a potential tragedy and genetic tyranny".

Appendix A

ORGANIC SEED SAVING

By Tom Stearns

People save seeds for different reasons: greater self-sufficiency, to maintain genetically and culturally valuable heirlooms, for fun, and to save money. The benefits obviously go far beyond these, and I hope to excite and motivate you to save your own seeds so that you too will get to experience this rewarding cycle.

Before saving the seed from the plants you are growing, there is some important planning that will save you time and frustration in your efforts, in addition to making them much more successful. First you need to decide what to grow. There are over 20,000 open-pollinated vegetable varieties available in the United States today. Hundreds of new varieties are bred each year, just as hundreds of old varieties are lost or rediscovered. The smaller and more regionally based seed companies will usually have more interesting varieties that are capable of producing seed in your growing season. With so many to choose from it is always best to pick your favourites and those that are easy to start with (beans, corn, lettuce).

The second important consideration when planning to save seed from your garden involves how the plants are pollinated and how far you might have to isolate members of the same species to ensure varietal purity. The major pollinators are wind and insects. Many plants are also self-pollinating, and therefore need very little isolation. See the vegetable list for the appropriate isolation for each crop.

Third, and most often overlooked, is how many individual plants to grow of each variety you are saving seed from. In some cases it does not take more than a few plants to give you enough seed, but you could be greatly risking the future genetic variability of your variety. Inbreeding depression is the term that is used for what happens when the loss of this variability results in a less vigorous and healthy plant. Corn is especially susceptible, and interestingly, this crop is the very same one that most frequently displays hybrid vigour. Corn seed should be selected from no less than 200 plants to prevent inbreeding depression.

After these simple but important steps have been taken, you are really prepared to save seed! The techniques for processing different types of seeds are relatively well studied and understood. However, it is always important and useful to watch what would happen to the seed itself if left on the plant until it fell. This can give major clues into how the seed needs to be processed in order to be stored for later planting. The plants are indeed our best teachers.

The different methods of processing most common vegetable seeds are grouped into two categories: wet seeds and dry seeds.

WET SEEDS
Squash, cucumbers, melons, and tomatoes
In most cases, these crops need to go through a fermentation process which helps them "burn off" any seed-borne diseases as well as prepare them for storage. This process is a common one for all of these crops, after a few preparatory steps have been taken. Both summer squash and cucumbers need to be left on the plant for longer than normal (4-6 weeks); once picked, they have many more mature seeds if left to sit inside for 4-8 weeks before being

Appendix A: Organic Seed Saving

fermented. Winter squash can be picked at the normal time and is best left inside for several months before being processed. Cucumbers should be left on the plant until their skins begin to turn orange and wrinkle slightly. Tomatoes should be allowed to ripen as much as possible – not to the point of rotting, but a little past the point where you would want to eat them. Once the squash, cucumbers and tomatoes have reached this point, the process of extracting the seeds is essentially the same.

Squeeze or scoop out the seeds and pulp, and mix this in a jar with about half as much water. When kept at room temperature, this mixture should start fermenting within two days. Stir every day to make sure that mold doesn't take over. In 4-5 days the good seeds should settle, while the immature seeds and pulp should float to the top. Pour off the scum, rinse the good seeds, and place them on a surface where they will dry within a day or two. If this process sounds a little complex, don't let it deter you; it is actually quite forgiving.

DRY SEEDS
Beans, peas, corn, lettuce, most flowers, radishes, and onions

These seeds are simply left to dry on the plant for as long as possible.

Some seasons are so wet that your seed will rot if not harvested before it is completely dry. The seed is getting its last bits of information from the mother plant at the end, so if it must be pulled, pull the entire plant with the seeds still attached so that they can mature as much as possible.

All seeds should be thoroughly dried and kept at temperatures below 40 degrees. Some crops will last for many years if kept this way, whereas some will last for only a few years. It is always best to use the freshest possible seed.

Remember also what I like to call the power of selection.

Changing the Nature of Nature

When you select to save seed from one individual plant over another you are making a powerful decision. You can either do it haphazardly, or you can improve varieties, with a great deal of intention, to work best under your conditions in the garden and in the kitchen.

SPECIFICS ON SAVING SEED FOR COMMON VEGETABLES

Corn-(sweet, flint, flour, dent, or pop)
wind pollinated, isolate varieties by 1/2 mile, grow at least 200 plants to prevent inbreeding depression, allow to dry on stalk, shell when completely dry.
Lettuce-self-pollinator, isolate varieties by 12 feet, allow plants to "bolt" and collect dry seed from maturing stalk.
Squash-insect-pollinated, isolate by 1/2 mile
Cucumbers-insect pollinated, isolate by 1/2 mile
Beans-(snap, dry, bush, or pole) self-pollinating, isolate by 15 feet to be sure of total purity (pole beans isolate 100 feet), allow to dry fully on plant, watch out for molding.
Peas-self-pollinator, isolate varieties by 5 feet for purity, allow to dry on vines but watch out for mold under moist conditions.
Melons-insect pollinated, isolate varieties by 1/2 mile
Tomatoes-self-pollinator, isolate varieties by 15 feet
Source: California Fresh Produce Guide, Publication No. 28 (Los Angeles: California Department of Food and Agricultures)

APPENDIX B

RELATED WEB SITES

Activist Network Squall www.squall.co.uk/ind2.html

Alive: Canadian Journal of Health and Nutrition www.concentric.net/~Rwolfson/home.html

Australian Gene Ethics Network www.zero.com.au/agen/

Ben & Jerry's www.benjerry.com

Biosafety Bibliographical Database www.icgeb.trieste.it/biosafety/bsfrel.htm

Biosafety Information Network and Advisory Service (BINAS) (service of the United Nations Industrial Development Organization) binas.unido.org/binas/

Biotechnology Industry Organization www.bio.org-welcome.html

British Natural Law Party Manifesto on GE www.natural-law-party.org.uk

Calgene www.calgene.com

California Food Policy Advocates www.cfpa.net

Californians for Pesticide Reform (CPR) www.igc.org/cpr

Campaign for Food Safety www.purefood.org

Campaign to Label GE Food www.thecampaign.org

Canada's Natural Law Party on GE www.natural-

Changing the Nature of Nature

law.ca/genetic/geindex.html

Center for Ethics and Toxics www.cetos.org

Certified Organic Food www.gks.com

Committee for Sustainable Agriculture www.csa-efc.org

Consumers International www.consumersinternational.org/campaigns/codex/

Consumers Choice Council consumerscouncil.org

Cooperative Resource from Unido and OECD www.oecd.org/ehs/biobin/

Corporate Watch www.corpwatch.org/

Council for Responsible Genetics www.gene-watch.org

Department of the Enviroment Genetically Modified Organisms (GMO) Information www.shef.ac.uk/~doe/

Earth First! Toxic Mutants www.k2net.co.uk/~savage/ef/

Farm Aid www.ervic.com/temp/farmaid/home.html

Food Research and Action Center www.frac.org

Foundation on Economic Trends www.biotechcentury.org

Free Range Activism Web Site www.gn.apc.org/pmhp/dc/genetics

Exposé of the rBGH Scandal www.notmilk.com

Food First www.foodfirst.org

Fox BGH Story www.foxbghsuit.com

'Franken-Food' News www.dorway.com/franken.html

Friends of the Earth Anti-GE Campaign www.foe.co.uk/camps/foodbio/genetic.html

Frost & Sullivan, Pro-Biotech Marketing, NY www.frost.com/

Genetic Engineering and Its Dangers.Dr Ron Epstein, SFSU userwww.sfsu.edu/~rone/gedanger.htm

Genetic Engineering Network www.dmac.co.uk/gen.html

Appendix B: Related Web Sites

Genetic Engineering Summary.
www.wholefoods.com/wfm/healthinfo/ bioengineering.html

Genetic ID www.genetic-id.com

Genetically Manipulated Food News. GE Archive
home.intekom.com/tm_info

GRAIN www.grain.org

Greenpeace Interntaional
www.greenpeace.org/~geneng/main.html

Greenpeace USA www.greenpeaceusa.org

Henry A. Wallace Institute for Alternative Agriculture
www.hawiaa.org

Hexterminators www.artactivist.com

Humane Farming Association www.hfa.org Indigenous Peoples Coalition Against Biopiracy www.niec.net/ipcd

Institute for Agriculture and Trade Policy www.iatp.org

International Center for Technology Assessment www.icta.org

ISAAA www.isaaa.org/

Monsanto www.monsanto.com

Monsanto's information on 'Roundup Ready' crops
www.roundupready.com/Soybeans/

Mothers and Others www.igc.apc.org/mothers/

Mothers for Natural Law www.safe-food.org

National Campaign for Sustainable Agriculture
www.sustainableagriculture.net

National Farmers Union www.nfu.org

National Organic Program ams.usda.gov/nop

Natural Law Party (US) www.natural-law.org/index.html

Novartis www.novartis.com

159

Organic Trade Association www.ota.com

PANOS www.oneworld.org/panos

Pesticide Action Network panna.org

Public Citizen www.citizen.org/

Pure Food Campaign www.purefood.org

Rachel's Environment & Health Weekly
www.monitor.net/rachel/

Rural Advancement Foundation International (RAFI)
www.rafi.org

Rural Vermont
www.greenfornewengland.org/members/rural.html

Summaries of Biotechnology, Genetic Engineering;
www.ucsusa.org/agriculture/biotech.html

Super Heroes Against Genetix
www.envirolink.org/orgs/shag/genetix.html

Sustainable Agriculture Network www.sare.org/san/

Swedish Physicians and Scientists Against G.E.
home1.swipnet.se/~w-18472/indexeng.htm

The FOX BGH TV Trial www.foxBGHsuit.com

Third World Network www.twnside.org.sg/souths/twn/bio.htm

UK Genetics Forum www.geneticsforum.org.uk/navbar.htm

UK list of gm free foods www.safe-food.org

UK Soil Association www.earthfoods.co.uk/soil.whator.html

Union of Concerned Scientists www.ucsusa.org

Vegetarian Society on Gen/Eng
catless.ncl.ac.uk/veg/Orgs/VegSocUK/Campaign/ genetics.html

Washington Biotechnology Action Council (WashBAC)
weber.u.washington.edu/~radin/

Whole Foods Market www.wholefoods.com

Appendix B: Related Web Sites

WWW Virtual Library on Biotechnology
www.cato.com/biotech/

APPENDIX C

ORGANISATIONS (US)

American Agricultural Movement 100 Maryland Avenue NE Box 69 Washington, DC 20002 Phone: 202-544-5750

American Council on Consumer Interests Anita Metzen, Executive Director 240 Stanley Hall University of Missouri Columbia, MO 65211 Phone: 314-882-3817 Fax: 314-884-4807

Associates of the National Agricultural Library Mark Andrews, Chairman 1555 Connecticut Ave., NW Suite 2000 Washington, DC 20006

California Food Policy Advocates 116 New Montgomery St. Suite 530 San Francisco, CA 94150 Phone: 802-254-1234 E-mail: cfpa@earthlink.net

Center for Environmental Information, Inc. 50 W. Main Street Rochester, NY 14614 Phone: 716-262-2870

Center for Ethics and Toxics P.O. Box 673 39141 South Highway One Gualala, CA 95445 Phone: 707-884-1700 Fax: 707-884-1846 E-mail: cetos@cetos.org

Committee for Sustainable Agriculture 406 Main St., #313 Watsonville, CA 95076 Phone: 831-763-2111 Fax: 831-763-2112 E-mail: csaefc@csa-efc.org

Congressional Legislation Hotline Phone: 202-225-1772

Consultative Group on International Agricultural Research (CGIAR) 1818 H Street, NW Washington, DC 20433 Phone: 202-

477-1234

Consumers Choice Council 200 P Street, NW, Suite 308 Washington, DC 20036 Phone: 202-785-1950 Fax: 202-452-9640 E-mail: consumer@ibm.net

Council for Responsible Genetics Martin Teitel, Ph.D. 5 Upland Road Cambridge, MA 02140 Phone: 617-868-0870 Fax: 617-491-5344 E-mail: crg@gene-watch.org

Earth Ethics Research Group, Inc. George Bortnyk, President 13938 85 Terrace North Seminole, FL 34646 Phone: 813-397-9042

Earthwatch P.O. Box 403 Mt. Auburn Street Watertown, MA 02272

Edmonds Institute Beth Burrows 29319-92nd Avenue West Edmonds, WA 98020 Phone: 425-775-5383 Fax: 425-670-8410 E-mail: beb@igc.apc.org

Environmental Research Foundation Peter Montague P.O. Box 5036 Annapolis, MD 21403 Phone: 410-263-1584 Fax: 410-263-8944 E-mail: erf@rachel.clark.net

Farm Aid Carolyn Mugar/Harry Smith 334 Broadway, Suite 5 Cambridge, MA 02139 Phone: 617-354-2922 Fax: 617-354-6992 E-mail: farmaid1@aol.com

Food and Water Phone: 800-EAT-SAFE Food Research and Action Center 1875 Connecticut Avenue, NW, Suite 540 Washington, DC 20009 Phone: 202-986-2200

Foundation on Economic Trends Jeremy Rifkin 1660 L Street NW Washington, DC 20036 Phone: 202-466-2823 Fax: 202-429-9602 E-mail: campaign@igc.apc.org

Friends of the Earth Brent Blackwelder 1025 Vermont Av NW Third Floor Washington, DC 20005-6303 Phone: 202-783-7400 Fax: 202-783-0444 E-mail: foe@foe.org Global Green USA. P.O. Box 21451 Columbus, OH 43221-0451 Phone: 805-565-3485

Global Response: Environmental Action Network P.O. Box 7490 Boulder, CO 80306-7490 Phone: 303-444-0306

Appendix C: Organisations (US)

Greenpeace USA Charles Margulis 1817 Gough Street Baltimore, MD 21231 Phone: 410-327-3770 Fax: 410-327-2990

Henry A. Wallace Institute for Alternative Agriculture 9200 Edmonston Road, Suite 117 Greenbelt, MD 20770-1551 Phone: 301-441-8777

High Mowing Organic Farm Tom Stearns 813 Brook Road Wolcott, VT 05680 Phone: 802-888-2480

Indian Law Resource Center Robert (Tim) Coulter 602 North Ewing Street Helena, MT 50601 Phone: 406-449-2006 Fax: 406-449-2031 E-mail: ilrc@mt.net

Indigenous Environmental Network Tom Goldsmith P.O. Box 485 Bemidji, MN 566601 Phone: 218-751-4967 Fax: 218-751-0561 E-mail: ien@igc.apc.org

Indigenous Peoples Coalition Against Biopiracy Debra Harry P.O. Box 72 Nixon, NV 89424 Phone: 702-574-0248 Fax: 702-574-0259 E-mail: ipcd@earthlink.net

Institute for Consumer Responsibility 5606 28th Avenue, NE Seattle, WA 98115 Phone: 206-523-0421

International Center for Technology Assessment Andrew Kimbrell 310 D Street NE Washington, DC 20002 Phone: 202-547-9359 Fax: 202-547-9429 E-mail: office@icta.org

International Forum on Globalization Victor Menotti P.O. Box 12218 San Francisco, CA 94112-0218 Phone: 415-771-3394 Fax: 415-771-1121 E-mail: ifg@igc.org

International Eco-Agriculture Technology Association, Inc. (IEATA) P.O. Box 998 Welches, OR 97067 Phone: 800-798-5543

Institute for Food and Development Policy: Food First! Peter Rosset 398 60th Street Oakland, CA 94618 Phone: 510-654-4400

Mothers and Others and Rural Vermont List of Organic and rBGH-Free Dairy Foods Phone: 888-ECO-INFO Phone: 802-233-7222

National Coalition Against the Misuse of Pesticides 701 E

Changing the Nature of Nature

Street, SE, Suite 200 Washington, DC 20003 Phone: 202-543-5450 E-mail: ncamp@igc.apc.org

National Farmers Union 400 Virginia Avenue SW Suite 710 Washington, DC 20024 Phone: 202-554-1600 E-mail: nfu@aol.com

Native Seed/Search Suzanne Nelson 2509 N. Campbell Suite 325 Tucson, AZ 85719 Phone: 520-327-9123 Fax: 520-327-5821

Organic Trade Association 50 Miles St. PO Box 1078 Greenfield, MA 01302 Phone: 413-774-7511 Fax: 413-774-6432 E-mail: ota@igc.org

Pesticide Action Network Ellen Hickey 116 New Montgomery Suite 810 San Francisco, CA 94015 Phone: 415-541-9140 Fax: 415-541-9253 E-mail: ehickey@panna.org

Pesticide Watch 450 Geary Street, Suite 500 San Francisco, CA 94102 Phone: 415-292-1486 E-mail: pestiwatch@igc.org

Pure Food Campaign Ronnie Cummins 860 Highway 61 Little Marais, MN 55614 Phone: 218-226-4164 Fax: 218-226-4164 E-mail: purefood@aol.com

Rural Advancement Foundation International-US Hope Shand P.O. Box 640 Pittsboro, NC 27312 Phone: 919-542-1396 Fax: 919-542-0069 E-mail: hope@rafiusa.org

Rural Vermont Ellen Taggart 15 Barre Street Montpelier, VT 05602 Phone: 802-223-7222 Fax: 802-223-0269 E-mail: rural-vermont@essential.org

Seed Savers Exchange Kent Whealy R.R. 3, Box 239 Decorah, IA 52101 Phone: 319-382-5990 Fax: 319-382-5872

Seeds of Change 621 Old Santa Fe Trail No. 10 Santa FE, NM 87501 Phone: 505-983-8956

Union of Concerned Scientists Margaret Mellon 1616 P Street NW, Suite 310 Washington, DC 20036 Phone: 202-332-0900 Fax: 202-332-0908 E-mail: mmellon@ucsusa.org

US Department of Agiculture 14th St. and Independence Ave.

Appendix C: Organisations (US)

SW Washington, DC 20250 Phone: 202-720-8732

US Department of Health, Food and Drug Administration (FDA) 5600 Fishers Lane Rockville, MD 20857 Phone: 301-443-1544

US Public Interest Research Group 215 Pennsylvania Avenue, SE Washington, DC 20003 Phone: 202-546-9707

Washington Biotechnology Action Council Phil Bereano Department of Technical Communication University of Washington 14 Loew Hall Box 352195 Seattle, WA 98195 Phone: 206-543-9037 Fax: 206-543-8858 E-mail: phil@uwtc.washington.edu

White House Office of Environmental Policy Old Executive Office Building, Room 360 1600 Pennsylvania Avenue Washington, DC 20501 Phone: 202-456-6224

Women's National Farm and Garden Association, Inc. Mrs. William Slattery, President P.O. Box 608 Northville, MI 48167 Phone: 419-422-2466

SUGGESTED READING

Aldridge, S., *The Thread of Life*: The Story of Genes and Genetic Engineering, Cambridge University Press, New York, 1996.

Anderson, R., Levy, E., and Morrison, B., Rice *Science and Development Politics*: IRRI's Strategies and Asian Diversity 1950?1980, Clarendon Press, Oxford, 1991.

Ausubel, K., *Seeds of Change: The Living Treasure; The Passionate Story of the Growing Movement to Restore Biodiversity and Revolutionize the Way We Think About Food*, Harper Collins, New York, 1994.

Bains, W., *Biotechnology from A to Z*, 2nd edition, Oxford Press, Oxford 1998.

Becker, G. K., ed., *Changing Nature's Course: The Ethical Challenge of Biotechnology*, Hong Kong University Press, Hong Kong, 1996.

Berry, W., *Home Economics*, North Point Press, San Francisco, 1987.

Biotechnology Industry Organization, *Agricultural*

Biotechnology: The Future of the World's Food Supply, Biotechnology Industry Organization, Washington, DC, 1995.

Boyce Thompson Institute, *Regulatory Considerations: Genetically Engineered Plants*, Center for Science Information, San Francisco, 1987.

Bradlow, Fishman, and Osborn, eds., *Cancer: Genetics and the Environment*, Annals of the NY Academy of Sciences, New York, 1997.

Bud, R., *The Uses of Life: A History of Biotechnology*, Cambridge University Press, Cambridge, MA, 1993.

Busch, L., Lacy, W.B., Burkhardt, J., and Lacy, L.R., Plants, *Power and Profit: Social, Economic and Ethical Consequences of the New Biotechnologies*, Basil Blackwell, Oxford, 1991.

Caseley, J.C., Cussans, G.W., and Atkins, R.K., *Herbicide-Resistant Weeds and Crops*, Butterworth-Heinemann, Oxford, 1991.

The Center for Public Integrity, *Unreasonable Risk: Politics of Pesticides*, Center for Public Integrity, Washington DC, 1998.

Cooper, I.P., *Biotechnology and the Law*, Clark Boardman Co., New York, 1989.

Council for Agricultural Science and Technology, *Herbicide Resistant Crops*, Council for Agricultural Science and Technology, Ames, IA, 1991.

Suggested Reading

Crawley, M., *The Biosafety Results of Field Tests of Genetically Modified Plants and Microorganisms*, Biologische Bundensanstalt für Land-und Fortwirtschaft, Braunschweig, 1992.

The Crucible Group, *People, Plants and Patients: The Impact of Intellectual Property on Trade, Plant Biodiversity, and Rural Society*, International Development Research Centre, Ottawa, 1994.

Darwin, C.R., *The Origin of Species by Means of Natural Selection or the Preservation of Favorable Races in the Struggle for Life*, Penguin, Harmondsworth, 1968.

Dawkins, R., *The Selfish Gene*, Oxford University Press, Oxford, 1976.

Dayan, A.D., Campbell, P.N., and Jukes, T.H., eds., *Hazards of Biotechnology: Real or Imaginary*, Proceedings of the Biological Council's Symposium in London, December 14?15, 1987, Elsevier, New York, 1989.

Doebley, J., *Proceedings of a USDA/EPA/FDA Transgenic Plant Conference*, September 7?9, 1988, Annapolis, MD, 1988.

Doyl, J., *Altered Harvest*, Viking, New York, 1985.

Duhon, D., *One Circle: How to Grow a Complete Diet in Less than 1000 Square Feet*, Ecological Action, Willits, CA, 1985.

Durning, A., *How Much is Enough? The Consumer Society and the Future of the Earth*, W.W. Norton, New York, 1992., *This Place on Earth: Home and the Practice of*

Permanence, Sasquatch Books, Seattle, 1996.

Elkington, J., *The Green Consumer*, Penguin Books, New York, 1990.

Ellis, B.W., ed., *The Organic Handbook of Natural Insect and Disease Control*, Rodale, Emmaus, PA, 1992.

Fowler, C., *Unnatural Selection: Technology, Politics and Plant Evolution*, Gordon and Breach, Amsterdam, 1994.

Fowler, C., and Mooney, P., *Shattering: Food, Politics and the Loss of Genetic Diversity*, University of Arizona Press, Tucson, AZ, 1990.

Frossard, P., *The Lottery of Life: The New Genetics and the Future of Mankind*, Transworld, London, 1991.

Goodman, D., Sorj, B., and Wilkinson, J., *From Farming to Biotechnology: A Theory of Agro-Industrial Development*, Iowa State University Press, Ames, IA, 1990.

Grace, E., *Biotechnology Unzipped*, Trifolium Books, Toronto, 1997.

Grossmann, E., and Atkinson, D., eds., *The Herbicide Glyphosphate*, Butterworth, London, 1985.

Harding, S., *The Science Question in Feminism*, Cornell University Press, Ithaca, NY, 1986.

Hawkes, J., *The Diversity of Crop Plants*, Harvard University Press, Cambridge, MA, 1983.

Suggested Reading

Ho, M.-W., *Genetic Engineering - Dream or Nightmare?: The Brave New World of Bad Science and Big Business*, Gateway Books, Bath, UK, 1998.

Hobbelink, H., *Biotechnology and the Future of World Agriculture*, Zed Books, London, 1991.

Holdrege, C., *Genetics and the Manipulation of Life: The Forgotten Factor of Context*, Lindisfarne Press, Hudson, NY, 1996.

Hubbard, R., and Wald, E., *Exploding the Gene Myth*, Beacon Press, Boston, 1993.
International Service for the Acquisition of Agri-Biotech Applications, *Global Review of Commercialized Transgenic Crops*, 1998.

Juma, C., *The Gene Hunters: Biotechnology and the Scramble for Seeds*, Cambridge University Press, Cambridge, 1989.

Kaiser, M., and Welin, S., eds., *Ethical Aspects of Modern Biotechnology*, Centre for Research Ethics, Goeteborg, 1995.

Kareiva, P., Manasse, R., and Morris, W., *Biological Monitoring of Genetically Engineered Plants and Microbes*, Agricultural Research Institute, Bethesda, MD, 1991.

Keller, E.F., *A Feeling for the Organism: The Life and Work of Barbara McClintock*, Freeman, NY, 1983., *Refiguring Life: Metaphors of Twentieth-Century Biology*, Columbia
University Press, New York, 1995., *Reflections on Gender and Science*, Yale University Press, New Haven, CT,

1985., *Secrets of Life, Secrets of Death: Essays on Language, Gender and Science*, Routledge, New York, 1992.

Kidd, J.S., and Kidd, R., *Life Lines: The Story of New Genetics*, Facts on File, New York, 1999.

Kloppenburg, J.R., Jr., *First the Seed: The Political Economy of Plant Biotechnology*, Cambridge University Press, Cambridge, MA, 1988.

Kneen, B., *Farmageddon: Food and the Culture of Biotechnology*, New Society Publishers, Gabriola Island, BC, 1999., *The Rape of Canola*, NC Press, Toronto, 1993.

Krebs, A.V., *The Corporate Reapers: A Book of Agribusiness*, Essential Books, Washington, DC, 1992.

Krimsky, S., *Bioethics and Society: The Rise of Industrial Genetics*, Praeger, New York, 1991.

Krimsky, S., and Wrubel, R., *Agricultural Biotechnology: An Environmental Outlook*, Department of Urban and Environmental Policy, Tufts University, Medford, MA, 1993., *Agricultural Biotechnology and the Environment*, University of Illinois Press, Urbana, IL, 1996.

Kung, S.D., and Wu, R., *Transgenic Plants, Vol. 1: Engineering and Utilization*, Academic Press, New York, 1993., *Transgenic Plants, Vol. 2: Present Status and Social and Economic Impacts*, Academic Press, New York, 1993.

Ladurie, R., *Times of Feast, Times of Famine: A History of Climate Since the Year 1000*, trans., Farrar, Straus and Giroux, New York, 1971.

Suggested Reading

Lampkin, N., *Organic Farming*, Farming Press, Ipswich, Suffolk, 1990.

Lappé, M., *Broken Code: The Exploitation of DNA*, Sierra Club Books, San Francisco, 1989., *Evolutionary Medicine: Rethinking the Origins of Disease*, Sierra Club Books, San Francisco, 1994.

Lappé, M., and Bailey, B., *Against the Grain: Biotechnology and the Corporate Takeover of Your Food*, Common Courage Press, Monroe, ME, 1998.

Lappé, M., and Collins, J., *Food First*, Abacus, London, 1982.

Latour, B., *We Have Never Been Modern*, Harvard University Press, Cambridge, MA, 1993.
Lear, J., *Recombinant DNA: The Untold Story*, Crown, New York, 1978.

Levins, R., and Lewontin, R., *The Dialectical Biologist*, Harvard University Press, Cambridge, MA, 1985.

Lewin, R., *Patterns in Evolution: The New Molecular View*, Scientific American Library, New York, 1997.

Lewontin, R.C., *Biology as Ideology: The Doctrine of DNA* (Massey Lectures), Anansi, Toronto, 1991.

Lewontin, R.C., Rose, S., and Kamin, L., *Not In Our Genes: Biology, Ideology and Human Nature*, Pantheon, New York, 1984.

Lycett, G.W., and Grierson, D., *Genetic Engineering of Crop Plants*, Butterworth, London, 1990.
Mander, J., I*n The Absence of the Sacred: The Failure of*

Technology and the Survival of the Indian Nations, Sierra Club Books, San Francisco, 1991.

Maranto, G., *Quest for Perfection: The Drive to Breed Better Human Beings*, Scribner, New York, 1996.

Mellon, M., and Rissler, J., eds, *Now or Never: Serious New Plans to Save a Natural Pest Control*, Union of Concerned Scientists, Washington DC, 1998.

Miles, M., and Shiva, V., *Ecofeminism*, Zed Books, London, 1993.

Millstone, E., *Food Additives: Taking the Lid Off What We Really Eat*, Penguin, Harmondsworth, 1986.

Molnar, J.J., and Kinnucan, H., eds., *Biotechnology and the New Agricultural Revolution*, Westview Press, Boulder, CO, 1989.

Morgan, D., *Merchants of Grain*, Research Foundation for Science, Technology and Natural Resource Policy, New Delhi, 1996.

Nabhan, G., *Enduring Seeds: Native American Agriculture and Wild Plant Conservation*, North Point Press, San Francisco, 1989.

Nicholl, D.S.T., *An Introduction to Genetic Engineering*, Cambridge University Press, New York, 1994.

Nottingham, S., *Eat Your Genes: How Genetically Modified Food is Entering Our Diet*, Zed Books, New York, 1998.
Office of Technology Assessment, *Commercial Biotechnology: An International Analysis*, US Government

Suggested Reading

Printing Office, Washington, DC, 1984., *New Developments in Biotechnology: Public Perceptions of Biotechnology*, US Government Printing Office, Washington, DC, 1987., *Technology, Public Policy and the Changing Structure of American Agriculture*, US Government Printing Office, Washington, DC, 1986.

Organization for Economic Cooperation and Development, *Biotechnology, Agriculture and Food*, 1992.

Pollack, R., *Signs of Life: The Language and Meaning of DNA*, Viking Penguin, Harmondsworth, 1994.
Raghavan, C., *Recolonisation: GATT, The Uruguay Round and the Third World*, Third World Network, Penang, Malaysia, 1990.

Rifkin, J., *The Biotech Century*, Tarcher/Putnam, New York, 1998., *Declaration of a Heretic*, Routledge Kegan Paul, Boston, 1985.

Rissler, J., and Mellon, M., *Ecological Risks of Engineered Crops*, MIT Press, Cambridge, MA, 1996., *Perils Amidst the Promise: Ecological Risks of Transgenic Crops in a Global Market*, Union of Concerned Scientists, Cambridge, MA, 1993.

Rose, S., *Lifelines: Biology Beyond Determinism*, Oxford University Press, Oxford, 1998.

Russo, E., and Cove, D., *Genetic Engineering: Dreams and Nightmares*, W.H. Freeman, New York, 1995.
Saign, G.C., *Green Essentials: What You Need to Know About the Environment*, Mercury House, San Francisco, 1994.

Shand, H., *Human Nature: Agricultural Biodiversity and Farm-Based Food Security*, The Rural Advancement Foundation International, Ottawa, 1997.

Shiva, V., *Biopiracy: The Plunder of Nature and Knowledge*, South End Press, Boston, 1997., *Globalization of Agriculture and the Growth of Food Insecurity*, Research Foundation for Science, Technology and Natural Resource Policy, New Delhi, 1996., *Monocultures of the Mind*, Zed Books, London, 1993., *The Violence of the Green Revolution*, Zed Books, London, 1991.

Shiva, V., and Moser, I., *Biopolitics: A Feminist and Ecological Reader on Biotechnology*, Zed Books, London, 1995.

Shiva, Ramprasad, Hegde, Krishnan, and Holla-Bhar, *The Seed Keepers*, Research Foundations for Science, Technology and Natural Resource Policy, New Delhi, 1995.

Shivanna, K.R., and Sawhney, V.K., eds., *Pollen Biotechnology for Crop Production and Improvement*, Cambridge University Press, New York, 1997.
Sofer, W.H., *Introduction to Genetic Engineering*, Butterworth-Heinemann, Boston, 1991.

Steinberg, M.L., and Cosloy, D.S., *Dictionary of Biotechnology and Genetic Engineering*, Facts on File, New York, 1994.

Steingraber, S., *Living Downstream: An Ecologist Looks at Cancer and the Environment*, Addison Wesley, Reading, MA, 1997.

Suggested Reading

Tagliaferro, L. *Genetic Engineering: Progress or Peril?*, Lerner Publications Co., 1997.

Teitel, M., *Rain Forest in Your Kitchen*, Island Press, Washington DC, 1992.

Tudge, C., *Food Crops for the Future*, Blackwell, Oxford, 1988.

Tudge, C., *The Engineer in the Garden: Genes and Genetics: From the Idea of Heredity to the Creation of Life*, Jonathan Cape, London, 1993.

Turney, J., *Frankenstein's Footsteps: Science, Genetics and Popular Culture*, St. Edmundsbury Press, Great Britain, 1998.

Vavilov, N.I., *The Origin, Variation, Immunity and Breeding of Cultivated Plants*, Roland Press, New York, 1951.

Vijayalakshmi, K., Radha, K.S., and Shiva, V., Neem: *A User's Manual,* Centre for Indian Knowledge Systems, Madras, and Research Foundation for Science, Technology and Natural Resource Policy, New Dehli, 1995.

Von Dommelen, A., ed., *Coping with Deliberate Release: The Limits of Risk Assessment*, International Centre for Human and Public Affairs, Tilburg, Netherlands, 1996.

Von Schomberg, R., ed., *Contested Technology: Ethics, Risks and Public Debate*, International Centre for Human and Public Affairs, Tilburg, Netherlands, 1995.

Watson, J.D., *The Double Helix: A Personal Account of the Discovery of the Structure of DNA*, Penguin, Harmondsworth, 1968.

Webb, T., and Lang, T., *Food Irradiation: The Myths and the Reality*, Thorsons, London, 1990.

Wheale, P., and McNally, R., eds., *The Bio-Revolution: Cornucopia or Pandora's Box?* Pluto Press, London, 1990.

Wills, C., *Children of Prometheus: The Accelerating Pace of Human Evolution*, Perseus Books, Reading, MA, 1998.

Winner, L., *The Whale and the Reactor: A Search for Limits in an Era of High Technology*, University of Chicago Press, Chicago, 1986.

Yoxen, E., *The Gene Business: Who Should Control Biology?*, Harper and Row, New York, 1983., *Unnatural Selection? Coming to Terms with the New Genetics*, Heinemann, London, 1986.

FOOTNOTES

CHAPTER ONE
1. California Agricultural Lands Project, *Genetic Engineering of Plants* (San Francisco: CALP, 1982), p. 13.
2. Michael Pollan, "Playing God in the Garden," *New York Times Sunday Magazine*, 25 October 1998.
3. Michael Antoniou, "GM Foods-Current Tests Are Inadequate Protection," *London Sunday Independent*, 21 February 1999.
4. Pollan, "Playing God in the Garden."
5. Correspondence between Roberto Verloza and Mark Ritchie, President, Institute for Agriculture and Trade Policy, 23 June 1999.
6. Brewster Kneen, *Farmageddon: Food and the Culture of Biotechnology* (Gabriola Island, BC: New Society Publishers, 1999), p. 206.
7. California Agricultural Lands Project, *Genetic Engineering of Plants*, p. 7.
8. Kneen, *Farmageddon*, p. 55.
9. Jack Doyle, *Altered Harvest: Agriculture, Genetics, and the Fate of the World's Food Supply* (New York: Viking, 1985), p. 177.
10. Ibid.
11. Kevin Bonham, "Biotechnology May Turn into Farmaceutical Delivery Systems," *Agweek Magazine*, 12 April 1999.
12. Clive James, "Global Review of Commercialized Transgenic Crops: 1998," International Service for the Acquisition of Agri-

Biotech Applications, Brief No. 8, p. 3.

CHAPTER TWO
1. Pollan, "Playing God in the Garden."
2. Council for Responsible Genetics, "Consumer Alert: FDA Approval of Flavr Savr Tomato Paves the Road for Genetically Engineered Foods," May 1994.
3. Sally Lehman, "Biotech Tomato Bruised," *San Francisco Examiner*, 10 January 1993.
4. Susan Benson and Leora Broyodo, "Flavor Saved?" *Mother Jones Magazine*, January/February 1997.
5. Lawrence Fisher, "Campbell Delays Plans on Biotech Tomatoes," *New York Times*, 12 January 1993.
6. Scott McMurray, "New Calgene Tomato Might Have Tasted Just as Good Without Genetic Alteration," *Wall Street Journal*, 12 January 1993.
7. Purdue University, "Genetically Engineered Food Already Plentiful in Basic American Diet," *Science Daily News Release*, 12 September 1998.
8. www.calgene.com/freshpr.htm.
9. Jonathan Friedland and Scott Kilman, "As Geneticists Develop Appetite for Greens, Mr. Romo Flourishes: Mexican Seed Billionaire Controls Seed Sales Amid Rush to Create New Strains," *Wall Street Journal*, 28 January 1999, sec. A.
10. Ibid.
11. Mae-Wan Ho, *Genetic Engineering-Dream or Nightmare? The Brave New World of Bad Science and Big Business* (Bath, UK: Gateway Books, 1998), p. 53.
12. Mae-Wan Ho, Hartmut Meyer, and Joe Cummins, "The Biotechnology Bubble," *The Ecologist*, Vol. 28, No. 3, May/June 1998, p. 149.
13. Michael Antoniou, "GM Foods: Current Tests Are Inadequate Protection," *London Sunday Independent*, 21 February 1999.
14. Stephen Nottingham, *Eat Your Genes* (New York: Zed Books, 1998), p. 76.
15. Bt spray is made from the fermented spores of the Bt bacteria

and has been in use for over fifty years.

16. Jane Kay, "Genetic Engineering v. Labelling Laws: A New Frontier," *San Francisco Examiner*, 11 July 1999.

17. R.P. Wrubel, S. Krimsky, et al. "Regulatory Oversight of Genetically Engineered Microorganisms: Has Regulation Inhibited Innovation?" *Journal of Environmental Management*, Vol. 21, No. 4, 1997, p. 578. www.tufts.edu/~skrimsky.

18. Ibid.

19. Pollan, "Playing God in the Garden"; and Wrubel, Krimsky, et al., "Regulatory Oversight of Genetically Engineered Microorganisms," pp. 571-586.

20. Kurt Schwartau, vice president of Agraquest, personal communication, 2 September 1999.

21. Margaret Mellon and Jane Rissler, "Now or Never: Serious New Plans to Save a Natural Pest Control," Union of Concerned Scientists, 1998.

22. David Holzman, "Agricultural Biotechnology," *Genetic Engineering News*, 15 April 1999, p. 12.

23. Ibid., p. 29.

24. Mellon and Rissler, "Now or Never"; and ISB Environmental Releases Database for the USDA APHIS web site: www.aphis.usda.gov/bbep/bp/index.html.

25. Marc Lappé and Britt Bailey, *Against the Grain* (Monroe, Maine: Common Courage Press, 1998), pp. 52-53.

26. Staff writer, "The X Fields," *Greenpeace Quarterly*, Winter 1996, pp. 9-11.

27. Ibid.

28. Caroline Cox, "Glyphosate (Roundup)," *Journal of Pesticide Reform*, Vol. 18, No. 3, Fall 1998, pp. 3-16.

29. RAFI has led the battle against the commercialization and proliferation of the terminator technology. For their extensive and thorough writing about the terminator and seed sterilization trends, visit www.rafi.org.

30. "Traitor Tech: The Terminator's Wider Implications," RAFI Communique, January/February 1999.

31. Council for Responsible Genetics letter to USDA, 15 October

1998.
32. Martha Crouch, "How the Terminator Terminates" (Edmonds, Wash.: The Edmonds Institute, 1998). Paper can be viewed at: http://www.bio.indiana.edu/people/terminator.html.
33. "Traitor Tech."
34. Jeff Kamen, "Formula for Disaster," *Penthouse*, March 1999.
35. Susan Gilbert, "Fears Over Milk, Long Dismissed, Still Simmer," *New York Times*, 19 January 1999; and the Harvard-based Physicians Health Study and Nurses Health Study.
36. Paul Kingsnorth, "Bovine Growth Hormones," *The Ecologist*, Vol. 28, No. 5, September/October 1998, p. 266.
37. Steven Gorelick, "Getting the Government on Your Side," *The Ecologist*, Vol. 28, No. 5, September/October 1998, p. 283.
38. To view the Canadian Health Reports visit: http://www.hc-sc.gc.ca/english/archives/rbst.
39. rBST "Gaps Analysis" Report by rBST Internal Review Team, Health Protection Branch, HealthCanada, 21 April 1998, p. 34.
40. Ibid., 35.
41. Ibid., 14.
42. Gilbert, "Fears Over Milk, Long Dismissed, Still Simmer."
43. Ibid.
44. Julie Nordlee et al., "Identification of a Brazil Nut Allergen in Transgenic Soybeans," *The New England Journal of Medicine*, Vol. 334, No. 11, 1996, pp. 688-692.
45. Warren Leary, "Genetic Engineering of Crops Can Spread Allergies Study Shows," *New York Times*, 14 March 1996, p. A20.
46. *Environmental Nutrition: The Newsletter of Diet, Nutrition and Health*, October 1996.
47. Ho, "The Biotechnology Bubble," p. 148.
48. Jaan Suurküla, "Bacteria with Kanamycin Resistance Genes Crossresistant to Valuable Antibiotics," Physicians and Scientists for Responsible Application of Science and Technology website: www.psrast.org/antibiot.htm.
49. Ho, *Genetic Engineering-Dream or Nightmare?*, p. 164.
50. Ibid., p. 154.
51. Greenpeace International Press Release, "Medical Experts

Footnotes

Consider Novartis Genetically Modified Maize Unacceptable," Brussels, 4 December 1998.
52. "Swiss Ban Genetically Modified Maize and Potatoes," Reuters, 17 April 1999.
53. Martha Crouch, "How the Terminator Terminates" (Edmonds, Wash.: The Edmonds Institute, 1998). Paper can be viewed at: http://www.bio.indiana.edu/people/terminator.html.
54. Ho, *Genetic Engineering-Dream or Nightmare?*, p. 165.
55. Hannelore Sudermann, "Green Genes," Spokesman Review, 16 August 1998, p. A15.
56. Ibid.
57. *Diamond v. Chakrabarty*, 100 S.Ct. 2204, 2206.
58. Petition is distributed by the Council for Responsible Genetics and is available at www.gene-watch.org.

CHAPTER THREE

1. "Foodborne Infections," Center for Disease Control website: www.cdc.gov/ncidod/dbmd/diseaseinfo/foodborneinfections_t.htm.
2. Nottingham, *Eat Your Genes,* p. 184.
3. Scott Allen, "Tinkering with the DNA on Your Dinner Plate," *Boston Globe*, 11 July 1999.
4. Antoniou, "GM Foods-Current Tests Are Inadequate Protection."
5. Carrie Swadener, "Bacillus Thuringiensis," *Journal of Pesticide Reform*, Vol. 14, No. 3, Fall 1994.
6. Ibid.
7. Fred Pearce and Debora Mackenzie, "It's Raining Pesticides," *New Scientist*, Vol. 162, No. 2180, 3 April 1999, p. 23.
8. Cox, "Glyphosate (Roundup)."
9. Ho, *Genetic Engineering-Dream or Nightmare?*, p. 148.
10. Genome Therapeutics Corporation, www.cric.com.
11. Ho, *Genetic Engineering-Dream or Nightmare?*, p. 148.
12. Debora MacKenzie, "Gut Reaction," New Scientist, 30 January 1999.
13. rBST "Gaps Analysis" Report, pp. 27-28.

14. Henry Miller, "A Rational Approach to Labelling Biotech Derived Foods," *Science*, Vol. 28, 28 May 1999, p. 1471.

CHAPTER FOUR
1. HRH the Prince of Wales, "Seeds of Disaster," *The Ecologist*, Vol. 28, No. 5, September/October 1998, pp. 252-253.
2. Gwynne Dyer, "Frankenstein Foods," *Toronto Globe and Mail*, 20 February 1999.
3. Jennifer Ferrara, "Revolving Doors: Monsanto and the Regulators," *The Ecologist*, Vol. 28, No. 5, September/October 1998, p. 285.
4. Paul Kingsworth, "Bovine Growth Hormones," *The Ecologist*, Vol. 28, No. 5, September/October 1998, p. 268.
5. Steven Gorelick, "Getting the Government on Your Side," p. 283.
6. Dyer, "Frankenstein Foods."
7. Mothers and Others 1999 list of rBGH-free dairy products: 888-ECO-INFO.
8. Press Release for Ben and Jerry's Homemade, Inc., "Legal Settlement Clears Way for National Anti-rBGH Label," 14 August 1997.
9. Ibid.
10. Mothers and Others and Rural Vermont list of organic and rBGH-free dairy foods: 888-ECO-INFO or 802-233-7222.
11. Ian Elliot, "Swedish Retailer Drops Ice Cream over Altered Soybeans," *Feedstuffs*, 2 June 1997.
12. 1998 Annual Report, Monsanto corporation.
13. 1998 Group Sales, Novartis Corporation.
14. 1998 DuPont Segment Sales & Net Assets.
15. Gillian Hadfield, "We Need a Label to Identify Genetically Altered Food," *Toronto Globe and Mail*, 10 May 1999, A15.
16. Bill Lambrecht, "US Turns Spotlight on Genetic Engineering," *St. Louis Post-Dispatch*, 30 May 1999.
17. Ho, Meyer, and Cummins, "The Biotechnology Bubble," p. 146.
18. Philip Bereano, "A Right to Know What You Eat," *GeneWatch*, Vol. 12, No. 1, February 1999, pp. 1-5.
19. Marian Burros, "Eating Well," *New York Times*, 21 May 1997, p.

C3.
20. Laura Eggerton, "Giant Food Companies Control Standards," *Toronto Star*, 28 April 1999.
21. www.bma.org.uk/public/science/genmod.htm.
22. Helen Gavaghan, "Britain Struggles to Turn Anti-GM Tide," *Science*, Vol. 284, May 1999, pp. 1442-1444.
23. Elliott, "Swedish Retailer Drops Ice Cream over Altered Soybeans."
24. Rick Weiss, "British Revolt Grows over 'Genetic' Foods," *Washington Post*, 29 April 1999, p. E2.
25. Ibid.
26. "Factbox: EU Consumer Responses on GM Food Issues," Reuters, 19 April 1999.
27. Peter Montague, "Biotech: The Pendulum Swings Back," *Peacework*, June 1999, p. 14.
28. The text of the bill can be viewed at: http://janus.state.me.us/legis/bills, and copies can be requested from www.state.me.us/legis.

CHAPTER FIVE
1. Jerry Mander, "How Cyber Culture Deletes Nature," *The Ecologist*, Vol. 29, No. 3, March/April 1999, p. 171.
2. Charles Benbrook, "Thin Air," *GeneWatch*, Vol. 12, No. 3, pp. 12-13.
3. Ibid.
4. Britt Bailey, Center for Ethics and Toxics, personal communication.
5. Burros, "Eating Well."
6. James Walsh, "Brave New Farm," *Time*, 11 January 1999.
7. Anuradha Mittal and Peter Rosset, "Seeds Sow Controversy," *San Francisco Chronicle*, 1 March 1999, p. A21.
8. Scott Kilman, "In New World of Tough Plants, Pesticide Sales Soften," *Wall Street Journal*, 16 June 1999, B6.
9. Cox, "Glyphosate (Roundup)."
10. "Monsanto Technology Agreement", Monsanto Company 1997
11. Rick Weiss, "Food War Claims Its Casualties as High-Tech Crop

Fight Victimizes Farmers," *Washington Post*, 12 September 1999, A7.
12. Melody Petersen, "Farmers' Right to Sue Grows, Raising Debate on Food Safety," *New York Times*, 1 June 1999, p. A1.
13. Ibid.
14. Paul Kingsnorth, "Is the Biotech Dream Crumbling?" *The Ecologist*, July 1999, p. 241.
15. Kamen, "Formula for Disaster."
16. Bill Lambrecht, "Interview with USDA Chief Glickman on the GE Controversy," *St. Louis Post-Dispatch*, 6 June 1999.
17. Scott Allen, "Group Lobbies for Labelling Genetically Altered Foods," *Boston Globe*, 18 June 1999, p. A3.

CHAPTER SIX
1. World Resources Institute, www.econet.apc.org/wri/biodiv/foodcrop.html.
2. Ho, *Genetic Engineering-Dream or Nightmare?* p. 124.
3. Alan Durning, *This Place on Earth: Home and the Practice of Permanence* (Seattle: Sasquatch Books, 1996), p. 7.
4. Ibid.
5. Solbring, p. 26.
6. Martin Teitel and Hope Shand, *The Ownership of Life: When Patents and Values Clash* (Minneapolis: Institute for Agriculture and Trade Policy, 1997), p. 11.
7. Vandana Shiva, Biopiracy: *The Plunder of Nature and Knowledge* (Boston: South End Press, 1997), p. 39.
8. Ibid., p. 55.
9. Ho, *Genetic Engineering-Dream or Nightmare?*, p. 125-6.
10. Hope Shand and the Rural Advancement Foundation International, *Human Nature: Agricultural Biodiversity and Farm-Based Food Security* (Ottawa, Ontario: Rural Advancement Foundation International, 1997), p. 22.
11. Seed Savers Exchange, *Garden Seed Inventory*, fifth ed. (Decorah, Iowa: Seed Savers Exchange, 1999).
12. Paul Waugh, "Official Data Reveals GM Crop Risks," *UK Independent*, 17 June 1999.

13. Vandana Shiva, *Betting on Biodiversity: Why Genetic Engineering Will Not Feed the Hungry* (New Delhi: Research Foundation for Science, Technology and Ecology, n.d.), p. 10; and Francesca Bray, "Agriculture for Developing Nations," *Scientific American*, July 1994, pp. 33-35.
14. Pollan, "Playing God in the Garden."
15. Scott Kilman, "Bugs May Develop Resistance to New Crops Faster than Expected," *Wall Street Journal*, 5 August 1999, p. A4.

CHAPTER SEVEN
1. *A Basic Call to Consciousness* (Mohawk Nation: Akwesasne Notes, 1978), p. 4.
2. 1999 *New York Times Almanac* (New York: Penguin, 1998), p. 488.
3. Genesis, 3:24.
4. For example, see Bruce Metzger and Michael D. Coogan, eds., *The Oxford Companion to the Bible* (New York: Oxford University Press, 1993), pp. 506-507.
5. Ibid., p. 507.
6. *A Basic Call to Consciousness*, p. 9.

CHAPTER EIGHT
1. Kevin Bonham, "Biotechnology May Turn into Farmaceutical Delivery Systems," *Agweek*, 12 April 1999.
2. David Holzman, "Monsanto Moves into the Contract Production Arena," *Genetic Engineering News*, Vol. 19, No. 4, 15 February 1999, p. 1.
3. Ibid., p. 8
4. Bonham, "Biotechnology May Turn into Farmaceutical Delivery Systems."
5. Ibid.
6. Ibid.
7. "GM Honey Could Help Medicine Go Down," Reuters, 23 June 1999.
8. Promar International, company press release, "Farmaceuticals: A New Era for Agriculture Food and Pharmaceuticals," 14 May

1999.
9. "Human Genes in Plants," *Halifax (Nova Scotia) Daily News* , 14 June 1999.
10. *All Things Considered*, National Public Radio, 16 March 1999.
11. International Food Information Council, "Food Biotechnology Benefits," www.ificinfo.health.org/brochure/biobenet.
12. Bonham, "Biotechnology May Turn into Farmaceutical Delivery Systems."
13. Gail Dutton, "Developments in Nutraceuticals," *Genetic Engineering News*, Vol. 19, No. 10, 15 May 1999, p. 59.
14. Carol Tacket et al., "Immunogenicity in Humans of a Recombinant Bacterial Antigen Delivered in a Transgenic Potato," *Nature Medicine*, Vol. 4, No. 5, May 1998, pp. 607-609.
15. J. Raloff, "Taters for Tots Provide an Edible Vaccine," *Science News Online*, 7 March 1998, www.sciencenews.org/sn_arc98/3_7_98/fob1.htm.
16. Anne Simon Moffat, "Engineering Plants to Cope with Metals," *Science*, Vol. 285, 16 July 1999, pp. 369-70.
17. US Center for Disease Control and Prevention, reported in *Food and Water*, Spring 1999, p. 10.
18. Dyer, "Frankenstein Foods."
19. Ibid.
20. Benny Haerlin, "Perils of Genetically Engineereed Crops," opinion, Reuters, 23 April 1999.

CHAPTER NINE
1. Seed Savers Exchange, *Garden Seed Inventory*.
2. Lucette Lagnado, "Strained Peace: Gerber Baby Food, Grilled by Greenpeace, Plans Swift Overhaul," *Wall Street Journal*, 30 July 1999, pp. A1, A6.
3. Weiss, "Food War Claims Its Casualties as High-Tech Crop Fight Victimizes Farmers."
4. Based on calculations using 1998 data, the most recent available. Scott Allen, "Tinkering with the DNA on Your Dinner Plate."I

INDEX

Against the Green. 82
AgrEvo 29
Agribusiness. 22, 28, 65-6, 74, 79
Akre, Jane 83
Allergy 38, 47-9, 110
Altieri, Professor Miguel. 99
American Cyanamid. 36
American Express. 90
American Soybean Association 50
Amikacin. 39
Amoxicillin 41
Ampicillin 40
Andes 101
Animal and Plant Health Inspection Service. 114
Antibiotic resistance 24, 39-42, 55-6, 101, 119
Antoniou, Michael 11, 52
Archer Daniels Midland 136
Argentina 53, 148
Artificial gene transfer 40
Asia. 47
Australia 53, 67, 120, 146, 148
Austria. 67
B.subtilis 39
Bacillus thuringiensis 26

Bailey, Britt	49-50, 82
BBC Television	142
Belgium	67, 74
Ben & Jerry's	68
Benbrook, Charles	77
Bereano, Philip	71, 92
Berlin	140
Bestfoods Inc.	72
Bill 291	33
Biodiversity	14, 96, 144
Bio-piracy	90, 93-4
Bio-pollution	39
Biosafety Protocol	148-9
Biotech	4
Biotech Baking Brigade	138
Biotechnology	8, 12-13, 92, 102, 148
Biotechnology Industry Organization	29
Birds Eye Wall's	73
Blair, Tony	141-4, 147, 149-50
BMA - British Medical Association	73
Booth, Cherie	143
Boycott, Rosie	146
Brazil	53, 121
BT - (Bacillus thuringiensis)	27-9, 39-40, 43, 52-4, 84, 97, 100-1
Calgene	7, 22-4, 43
California	32, 49, 132, 138
California Department of Food and Agricultures	156
California Fresh Produce Guide	156
Campbell's Soup Company	23
Canada	2, 25, 35-6, 53, 67, 69, 72, 97, 114-5, 148
Carbenicillin	41
Cargill	148-9
Carnation Alsoy	71
Carrefour	73
Catholics	107
CBC - The Canadian Broadcasting Corporation	35

Index

Center for Ethics and Toxics 49
Center for Food Safety 58, 84
CETOS 49-51, 53
CGIAR - Consultative Group on International Agricultural Research. . 33
Charles, Prince of Wales 61, 120, 145
Chile.. 132, 139, 148
China 53, 89, 99, 101
Christianity.. 108
Clinton, Bill 66, 143
Cloxacillin... 41
Codex Alimentarius Commission............................ 72
Common Courage Press 82
Consumers Choice Council 135
Control of Plant Gene Expression 32
Coopers Agropharm 36
Council for Responsible Genetics........................ 33, 135
Crick 18
Crouch, Dr. Martha.................................... 34, 41
CSA - Community Supported Agriculture 126
Daily Express... 146
Daily Mail 146
DeFelice, Dr. 116
DeKalb 16
Delhaize... 74
Delta & Pine Land 32, 39
Democratic Party 66
Denmark 67
Deutsche Bank... 148
Diamond vs. Chakrabarty 43
DNA Plant Technology..................................... 25
Dolly 11
Doritos.. 72
Dow 21
Downing Street... 147
DuPont 21, 28, 69, 114
East Germany.. 42

Eden 71
Edinburgh 149
Effelunga 74
Elanco/Eli Lilly 36
Empresas La Moderna 25, 44
Endless Summer Tomato 25
Enfamil Prosobee 71
Environmental Defense Fund 31, 37
Environmental Nutrition 38
EPA - US Environmental Protection Agency 28, 52-3, 64, 80, 84
Europe . 40, 54, 58, 65, 73-4, 94, 98, 120, 124-5, 136, 138-9, 143-4, 148
Exxon 80
Farmaceuticals and Pharming 115
FDA (Food and Drugs Administration) 21-2, 28, 36, 58-9, 65-8, 70, 84-5, 117
Fertilizers 99
Fidher, Linda 69
Finch, Sarah 148
Finland 67
Flavr Savr 22-4, 39, 43
Food Disparagement Laws 81
Ford 80, 86
Foudin, Arnold 114
Foundation on Economic Trends 84
Fowler, Cary 91
Fox TV 83
France 67, 73
Friends of the Earth 135, 144, 147
Fritos 72
Gandhi, Mahatma 138
Gaps Report 35-6, 57-8
Garden Seed Inventory 129
Genesis 109
Genetic Engineering News 116
Genetic erosion 17
Genetic ID 71

194

Index

GeneWatch 133
Genome Therapeutics Corporation 55
Gerber 78, 136
Germany 67, 143
Giddings, Val 29
Glickman, Dan 85
Glyphosate 32, 54, 80
GM foods (labelling)21, 48, 51, 58, 62-3, 65-6, 68, 71, 73-5, 78, 98, 125-7
GNA 141
Goldberg, Dr. Rebecca 37
Gore, Al 31
Great Britain . 2, 41, 46, 54, 61, 67, 73, 90, 107, 120, 126, 138, 141, 143
Greece 67, 106
Greenpeace 31, 84, 120, 136, 146-9, 151
Hadfield, Gillian K. 69
Haerlin, Benny 120
Harrington, Rev. Mary J. 106
Hatch, Senator Orrin 116
Havenaar, Robert 56
HealthCanada 35
Herbicide-resistance 30, 39, 52, 54
Highbeta-carotene oil 115
Hinduism 107
Hoover Institute 59
Horizontal gene transfer 38
House of Lords 146
Hybridisation 15-16, 18, 74
IGF-1 34-5
Illinois 68, 82
India 53, 74, 93, 99-100, 102, 149
Indian Gene Campaign 149
Indiana State University 34
Indiana University 40-1
Innovator canola 25
Intergrated Protein Technology 114
interleukin-10 115

International Society of Chemotherapy ... 41
International Union for the Conservation of Nature ... 120
Iolcos Grace ... 148-9
Iowa ... 71, 129
IPT ... 114
Ireland ... 16, 67, 74, 106
Irish potato famine ... 16, 95
Isomil ... 71
Israel ... 67
Italy ... 67, 74, 89
J. Sainsbury Plc ... 73
James, Professor Philip ... 142
Japan ... 53, 65, 106, 139
Jeavons, John ... 128
Judaism ... 107-8
John Innes Centre ... 98
Journal of Pesticide Reform ... 32
Juniper, Tony ... 144
Kanamycin ... 24, 39
Konsum ... 73
Krimsky, Professor Sheldon ... 27
Labour Party ... 146, 150
Lambrecht, Bill ... 70
Lancet magazine ... 34-5
Lappe, Mark ... 49-50, 82
Lindow, Stephen ... 26
Lukoskie, Luke ... 31
Luxembourg ... 67
MacGregor's ... 23
Mad cow disease ... 46, 49, 61, 147
Maine ... 74
Mander, Jerry ... 77
Marks and Spencer ... 73
Martini, Betty ... 67
Maryanski, James ... 65
May, Sir Robert ... 146-7

Index

MCPA 54
Meacher, Michael 143, 147-8
Mead Johnson & Co. 72
Melchett, Lord 146
Mellon, Margaret 28
Methicillin 41
Methionine 37
Mexico 23, 100, 139
Migros 74
Miller, Henry 59
Miller, Margaret 67
Ministry of Agriulture, Fisheries and Food 98
Minneapolis 103
Mombiot, George 150
Monoculture 16, 95-6
Monsanto11, 21, 24-5, 27-8, 30-2, 34-6, 38-9, 41, 46, 49-51, 54, 57-8, 64, 66-7, 69-70, 72,77-80, 82-3,114-5, 118, 120, 135, 138-9, 148-9
Montreal 147, 149
Morningstar Farms 72
Mothers & Others 68
Mowlam, Mo. 149-50
MSG - Monosodium Glutamate 47, 62
National Academy of Sciences 17
Native Americans 111
Nature Medicine 117
Neomycin 24
NeoRx 114
Nestlé 72-3, 123
Netherlands 56, 67, 115, 121
New England Journal of Medicine 37
New Hampshire 33-4
New Leaf Potato 28, 64, 70
New Leaf Product Guide and Seed Directory 64
New York 2
New York Times 36, 71
New Zealand 67

Nigeria	90
NIH	57
Norway	67
Novartis	33, 39-40, 69, 78-9, 136, 139
NutraSweet	62, 71
Nutrition Facts	63
Odak, Perry	68
OECD	149
Organic Valley	68
Orphan Drug Act	116
Paraguay	106
Penethicillin	41
Penicillin	40, 55
Penwells of Saltash	83
Pesticide	27, 52, 64, 99
Pesticide-resistance	39
Phytoestrogens	50-1, 58
Pioneer Hi-Bred	16, 37
Pollan, Michael	10, 11
Portugal	67
Posilac	57, 66
Princeton	147
Proceedings of the National Academy of Sciences USA	40
Proctor & Gamble	72
Promar International	115
Pseudomonas syringae	26
Pusztai, Arpad	53-4, 141-2
Rabideau, Marie	33
RAFI - The Rural Advancement Foundation International	32
rBGH - Recombinant Bovine Growth Hormone	34-7, 57-8, 66-9, 83-4
rBST - Bovine Somatotropin	34, 36, 58
Red Cross	107
Republican Party	66
Rifkin, Jeremy	84
Roundup Ready	30-2. 38, 50, 54, 80
Rowett Research Institute	53, 142

Index

Rural Advancement Foundation International 96
Rural Vermont 69
Saccharin 62
Sahai, Surman 149
Salmonella 56
Sao Paulo 103
Schmeiser, Percy 79
Science for the People 147
Sechen, Suzanne 67
Secrett, Charles 147
Seed Savers Exchange 97, 129
Seed Trade Act 94
Shapiro, Robert 72, 138
Shiva, Vandana 93-4
Similac Neocare 71
Six Nations Confederacy 105, 111
South Africa 53
South-East Asia 105
Spain 67
Spokesman Review 42
St. Louis Pist Dispatch 70
StarLink 29
Stonyfield Farms 68
Streptothricin 42
Sudermann, Hannelore 42
Superbugs 3, 29, 100-1
SuperQuinn 74
Superweeds 3, 39
Sweden 67, 73
Switzerland 41, 67, 74, 136
Taiwan 2
Taylor, Michael 67
Tennessee 101
Tesco 73
Tetracycline 41
Texas 81

The Daily Mail 145
The Ecologist 25, 70, 82-3, 133
The Guardian 143, 146
The Independent on Sunday 147
The Monsanto Files 83
The Nutritional Research and Education Act 116
The Washington Post 73
Thoreau, Henry David 138
Tibet 101
Time magazine 78
Titanic 113
TNO Nutrition and Food Research Institute 56
Tobramycin 39
Toronto Globe and Mail 66
Tositos Crispy Rounds 72
Turkey 101, 106
UK Government 141-2
The Impact of Genetic Modification on Agriculture, Food and Health. 73
UN Food and Agriculture Organization 72
Unilever UK 73
Union of Concerned Scientists 28
United Nations 107
United Nations Food and Agriculture Organization 91, 96
United Nations World Health Organization 72
United States 2, 18, 23-5, 33, 36-7, 46, 48, 51-3, 55, 57-8, 62-5, 67, 69, 71-2, 75, 78-9, 82, 85, 88, 91, 96-7, 99, 101, 106, 120-1, 124, 126-7, 129, 131-2, 136-40, 143-4, 146, 148-9, 153
University of Arizona 101
University of California, Berkeley 26, 99
University of Toronto 69
University of Washington, Seattle 71
Uruguay 148
US Agricultural Department 120
US Center for Disease Control and Prevention 119
US Congress 43, 85, 91, 116
US Department of Agriculture 85

Index

US Government 22, 36-7, 50, 57, 59, 70-1, 74
US Supreme Court 42, 81, 92
USDA - United States Department of Agriculture 18, 32-3 85, 114
Van der Bergh 73
Vegetative Propagation 15
Verloza, Roberto. 11
Vermont 66-8
Virus-resistance 25
Vital Health Publishing. 82
Vogel, Orville. 42
Wales. 148
Wall Street Journal 23
Washington DC. 31, 67, 85
Washington State University 42
Watson. 18
West Africa 105
Wheat rust epidemic 17
White, William 114
Whole Foods Market Inc. 68
Wilson, Steve. 83
Winfrey, Oprah 81-2
World Development Movement 148
World in Action 142
Zeneca. 79
Zurich 103

Vision Paperbacks publishes a wide range of exciting current affairs and investigative titles. If you would like to know more, or if you would like to order any of the books profiled in the following pages, you can call our orders hotline at:

Vision Paperbacks
20 Queen Anne Street
London W1M 0AY
UK

phone: +44 (0)20 7323 9757
fax: +44 (0)20 7323 9747

or email visionpaperbacks@compuserve.com

Other Titles from Vision Paperbacks

Car Wars
Battles on the Road to Nowhere
Chris Mosey

The car and the problems it causes in the modern world are portents of a fast gathering environmental holocaust that we are bringing upon ourselves.

Car Wars offers a provocative and challenging investigation into the invention and development of the car. In doing so, author Chris Mosey traces its turbulent history and the equally controversial developments of British transport policy.

Through a series of absorbing and revealing interviews, Mosey casts a critical eye at another modern phenomenon – the Eco-warrior. Considering the role of anarchist green groups such as Reclaim The Streets, he asks how effective their protests really are. One of the basic premises of *Car Wars* is that the problems caused by the car are more acute in Britain than elsewhere in the western world, and that it is here that they will have to be dealt with first. Other countries have the same problem, but they also have more space, which allows the ravages of so-called progress to be less visible. In crowded Britain, the blight on cities and countryside caused by the automobile and road-building is only too apparent.

'Progress' is presently leading us towards a global economy. This, our leaders tell us, will be our salvation. Yet a global economy without global government can only accelerate our destruction of the planet. It grants multi-national companies the de facto right to ride roughshod over environmental concerns, which usually start at a very local level, and to set countries like China and India travelling down the same road that we in the West are just beginning to realise has taken us in the wrong direction.

Mosey's sharp-witted analysis poses the question, can we effect the change needed to halt the environmental destruction of our planet? Or are or are we really heading down a road to nowhere?

£9.99
ISBN: 1-901250-40-7

Censored
The News They Don't Want You to Read
David Northmore

The role of the investigative journalist has never been so important. While alleged 'dumbing down' of newspapers and television programming intensifies and investigative budgets dwindle, open information is becoming an increasingly rare commodity. The forthcoming Freedom of Information Act reforms in the UK are highly unlikely to open the nation's official secrets to the public, despite high hopes and many journalists, lawyers and civil liberty campaigners have criticised the reforms as not being worth the paper they are printed on. *Censored: The News They Don't Want You to Read* is an outstanding collection of major journalistic investigations that challenge the 'dumbing-down' of the media and for the first time creates a focal point for important stories that one might not otherwise be researched.

Produced by reporters from the Association of Investigative Journalists, Censored blows the whistle on poisoned water supplies; MI5's secret computer files; incompetent coroners; death-trap cars; illegal council contracts; the new slum landlords; the 2.5 million tonne annual trade in asbestos; death by aerosol; condoms linked to cancer; Big Brother's sinister spy network; the secret sex lives of judges – and much more. In addition to these scoops, this book looks at the nature of censorship of investigative reporting and also includes an invaluable guide to the people, groups, websites and publications that strive to challenge censorship in the UK and elsewhere.

The first in an annual series of censored news stories, *Censored* represents a vital guide to community activists, lawyers, trade unionists, politicians and anyone else who believes that the public 'has a right to know'.

£9.99
ISBN: 1-901250-53-9

Other Titles from Vision Paperbacks

Do No Harm?
Munchausen By Proxy Syndrome
Craig McGill

Munchausen by Proxy Syndrome is heralded as a serious mental illness. It is said to prompt parents and care workers to inflict pain on their children, and its diagnosis leads children to be taken away from their parents and placed into care. And yet there is little, if any, evidence that the syndrome actually exists.

In *Do No Harm?* journalist Craig McGill claims that after unsubstantiated and badly researched diagnoses of MBPS, children are withdrawn from their family environment and placed into foster and care homes; causing trauma for both the children and the parents. Using specific case studies, as well as documented scientific evidence, McGill argues a vital case for the well-being of children across Britain and the rest of the world.

The book looks at the origins of MBPS – it 'arrived' from the US in 1951 – and examines the most public cases of the last 50 years. McGill goes on to discuss the situation today, looking at who is being accused, and why. *Do No Harm?* also examines the role the media has played in the propagation of MBPS. With so many media representations of the syndrome on television and in the newspapers, is the syndrome becoming mythologised into existence across the country?

Craig McGill examines the syndrome's appearance throughout the UK, Europe and the US, leading the reader to draw worrying conclusions about the validity of accusations of MBPS across the western world. *Do No Harm?* concludes with a section that examines what can be done to ameliorate the situation. Can European legislation work? Will the law see sense? Or will there be more and more mothers screaming for the return of their children without being heard?

£9.99
ISBN: 1-901250-048-2

Forgotten Children
The Secret Abuse Scandal in Children's Homes
Christian Wolmar

An extraordinary scourge swept across Britain – thousands of children taken into care by local authorities were abused by those charged with looking after them. This happened for almost 20years and little action was taken in this time. The legacy of this disaster is still being felt; major police investigations have been launched across the country and many of these are ongoing. Some compensation has been paid and the total bill will come to tens of millions of pounds but in the meantime a generation of children raised in care have had their lives blighted by abuse.

Why did it happen? Despite several official investigations, none have attempted to explain the underlying causes of institutional abuse, and *Forgotten Children* is the first book to attempt to do so. It charts the history of children's homes, how they were neglected over the years and why in the 1970s and 1980s they became preying grounds for paedophiles. It also shows how changes in social services provision in the 1970s helped to create the preconditions for disaster.

Christian Wolmar analyses the role of all the institutions that were found wanting in that they 'allowed' this scandal to sweep the country; local authorities and charities that allowed abuse to go unchecked in their own homes; central government, which failed to heed the warning signs; the police who initially ignored all complaints and the social workers who failed to listen to the children.

The legacy of the scandals goes well beyond its immediate victims, many of whom have become too traumatised to lead functional lives or have even committed suicide: prisons are full of the former residents of children's homes and their crimes have, in turn, created a whole raft of other victims.

£9.99
ISBN: 1-901250-47-4

Other Titles from Vision Paperbacks

Getting Out
Life Stories of Women who left Abusive Men
Ann Goetting

Each year, more than 50,000 women and children flee their homes as a result of domestic violence in the UK. Many women are forced to seek restraining orders on their partners in order to protect themselves. While many books detail distinguishing characteristics of the abusive relationship and the psychological profiles of the women and men involved, few accounts reveal how some women eventually find the courage to get up and leave.

In a harrowing yet inspiring chronicle, Ann Goetting in the US and Caroline Jory in the UK tell of the women who got away for good.

Getting Out recounts not only the stories of their abuse but also the women's life histories leading up to the violence – and the resources they drew upon to escape.

Some received assistance from family members or were saved by a network of friends, whilst others sought refuge in women's shelters.

The author explains that leaving is a process rather than an event, often marked along the way by reconciliation and the resumption of abuse. But as she and the interviewees suggest, the process invariably extends back to a critical moment when a decision is made to leave. Such a life-affirming moment may follow a particularly appalling episode of abuse. Other times it arrives in a long-repressed recognition of self worth gained from positive experiences at work or in parenting.

Getting Out is a book that will inspire and encourage women to discover solutions to problems within their own lives and those of the people they know. It is also a title that social workers and psychologists who deal with abused women will find informative. This work reaches out to an audience of readers seeking to understand the lives of women involved with abusive men.

£9.99
ISBN: 1-901250-51-2